The Ten Commandments For Life On The Road
© 2024, by Jon L. Baker

For the Nomad that lives deep within each of us, and all travelers everywhere.

The Ten Commandments For Life On The Road

I. You Are Never In A Hurry
Take you time and enjoy the ride.

II. Less Stuff = Less Stress
Life is better without!

III. Mind Your Resources
Use less. Need less.

IV. Keep It Simple
Nothing is perfect, and it doesn't have to be.

V. Maintain Thy Rig
Your house has wheels. Keep 'em rolling!

VI. Keep It Clean
Inside and out!

VII. Leave No Trace
Better than you found it.

VII. Pay It Forward
Give what you have. Get what you need.

IX. Thou Shalt Not Be A Dick
Self-explanatory.

X. You Can Always Drive Away
You don't have to attend every fight to which you're invited.

Foreword

The Road Is My Higher Power.

Despite the title, this book is not intended to be spiritual or religious, nor is it meant to mock anyone's belief system. We have a deep respect for people of faith and those that practice spirituality. We use the phrase "10 Commandments" merely as a lighthearted and convenient way to package and organize certain tenets that we have learned along our Nomadic Path.

With this book, we hope to paint a realistic picture for aspiring Skoolie/Van/RV-Dwellers of what life on The Road entails, what to expect, what to prioritize, and how to avoid the most common pitfalls. Our goal is to save you some time and trouble, as well as a lot of money and frustration.

That being said, "The Road is my higher power."

I've always known it. I get an almost inexplicable sense of comfort and purpose from being on The Road. - I belong there.

As a child I would daydream about traveling. I'd fantasize about watching the fields and towns endlessly passing by, with an insatiable urge to learn what was beyond the next horizon... and the next one... and the one after that... In my mind's eye I would see the signs and traffic signals, country roads and city streets, the hills, the valleys, the curves, buildings, bridges, phone poles, and dotted yellow lines. I could feel the cracks in the blacktop as The Road rushed by beneath me.

That was my happy place; being in motion, going somewhere, going anywhere, or going nowhere at all. It was the place where I felt "right". I was drawn to it. The destination wasn't important. It was The Road that mattered. It was The Road that called my name. The Road has always been my home.

At the age of five I decided that I was going to be a musician. I saw an episode of "The Monkees" (a 60s TV show about a goofy Rock Band, loosely based on the Beatles). I decided that this was my calling, and I never made a backup plan. All through childhood, I just knew that I was going to be a rock star. So... no need to consider anything else.

When I look back on it, however, my younger self would actually fantasize much more about being on The Road than I would about writing, recording, or playing sold-out shows to festival crowds. - I would think about being on tour more than I would about being in a band.

Though far from famous, I was very fortunate to eventually travel and play music professionally, and while I certainly loved the art, the creation, the camaraderie, and the applause, I did, as my younger self imagined, love being on The Road even more.

I've played some amazing shows, with amazing people, in amazing places, but in many ways, getting back on the bus after the lights were out and all the gear was loaded was even better. "Great gig guys... Off to the next town."

I also consider myself incredibly lucky to have grown up in an era in which roads ruled the day.

A little over a hundred years ago this was not the case. Sure, we had what the ancient Romans had built across Europe, and a mishmash grid of paved streets in most major cities, but for most of the world, what passed for "roads" were nothing more than wagon wheel ruts stretching across the plains, or a dirt path through the woods heading into the next town.

Our ancestors rode horses, or walked for days on end along unmarked trails, through rough country and mountain passes, facing whatever danger may come. They might have to leave home for weeks at a time just to make their way to the nearest public market.

A few generations ago people would spend 4-6 months on The Oregon Trail just to get halfway across the U.S., for nothing more than dreams of a better tomorrow. Today it can be done in 2 days (3 if you take the back roads, and enjoy the ride). Travel used to be an arduous and dangerous undertaking by foot, by horse, or by carriage. Today we have cars and airplanes, but more importantly, we have roads, available and waiting for you, anytime, day or night.

This, unfortunately, won't be the case 100 years in the future, either.

Our descendants, thanks to advances in technology, will probably just push a button to go hundreds, if not thousands of miles, from city to city in a relative instant. Even now, there are bullet trains that can get you from Tokyo to Osaka in a couple of hours (roughly the distance from St Louis to Nashville).

To our grandchildren, the path, the trail, the struggle, The Road... will be meaningless. The journey won't even be a consideration. You just push the button and you're there... That's it.

It is only this narrow group of so very few generations that have grown up with distinct, well designed, and arguably well maintained, asphalt and concrete paths, with clear and concise maps, giving specific instructions on how to get from here to <u>anywhere at all</u>.

It is amazing to me how many people take it for granted.

The Road, in itself, is the very epitome of the history of Western Society. It holds the secrets of the past in plain view, yet keeps them obscurely hidden. The subtle signs of yesterday are peppered among the layers of time that have built up along the way.

Old animal migrations and nomadic paths grew to be hunting trails, wagon train routes, farming roads, and local thoroughfares, all eventually leading to the highways and freeways of today. It has been a steady process of building and rebuilding, sparked by progress and dreams.

The Road was born of yearning. It was a part of that wanderlust that lies deep within us all. Sure, there's the necessity of having a simple way to make it into town, but The Road is so much more than that.

Along every step of the way there are traces of the people and cultures that came before. Stone bridges, faded signs, ramshackle houses, and abandoned roadside attractions dot the landscape. To the casual observer they may seem out of place, as if history somehow dropped them out of the blue, for no other reason than to spark our curiosity.

Some might even see them as eyesores, obsolete infrastructure to be removed and replaced by 7-11s and Walmarts. To others, they aren't even there at all. A shocking number of people don't even give them a second thought or a casual glance. They drive right by, phone in hand, thinking about that upcoming meeting, or what to pick up for dinner on the way home.

Yet, there they are, the legacy of our forefathers and beacons to times gone by. They stand as sentries, echoes of the past. They are run-down, dilapidated, and crumbling, but somehow immaculate and stunning in their representation of the lives and moments we've left behind.

Drive through any small town and you'll likely see an abandoned barn sitting next to a 20 year old strip mall that, itself, is about to be torn down to make room for a big-box store. How many of you have come across a group of log cabins, slowly rotting away next to the freeway, or a string of 50s-era motels with their elaborate neon signs still flickering in the night, flashing along a city street that no longer seems to connect to anything important? These are all indicators of what once was, a space in time in an ever-changing culture.

The history of The Road is thick and palpable. You can taste it. You can feel its' presence. It sings to you. It calls you. It takes you to yesterday, and it shows you tomorrow in the same instant.

It is ever-changing, yet it's completely and utterly constant.

The very soul of everyone who has traveled The Road before beckons to you, like a long lost love. The people who built those stone bridges and neon signs left behind a shadow of their hope, and images of their dreams. You can almost hear their voices. You can almost feel them, still moving along beside you, chasing that horizon.

Make no mistake. Roads are Magical, and the places they can take you are limitless. We mean this not just in a physical sense, but mentally and emotionally as well. The Road will take you anywhere you care to go, and any way you care to feel. Every place you've ever imagined, any person you've ever known, and whatever memories you care to create are somewhere out there, somewhere down The Road.

At the same time, The Road has no memory and no mercy. It holds no obligations, and makes no promises. *There is only right now*. All the past is forgiven. All the future is unwritten. There is <u>this</u> moment, <u>this</u> milepost, <u>this</u> gentle curve to the left, and nothing else. - Until you reach the next moment, and then the one after that.

Being on The Road brings me solace. It puts me at peace and makes me feel part of something bigger, something larger than myself and whatever drudgery the day may bring. It doesn't matter where I'm going, or why. It only matters that I'm on a path... any path at all. It's that feeling of moving on to better things. No matter how mundane the task, or how unpleasant the day, it is the motion. It is The Road.

Living on The Road is <u>freedom</u>. You can make whatever choice you want to make. Turn left. Turn right. Pull over at the next gas station. Stop for a moment to take in the view, or put the pedal down and push hard until morning. Drive for days to visit old friends, and make new ones along the way.

It can take you from the valley of the sun to snow crested mountains. It can take you from amber waves of grain to purple mountain majesty, and from crowded city streets to the middle of nowhere. You can have whatever you want, and be wherever you want to be. You just have to turn the key and put it in gear.

So, yes, The Road is my higher power.

I don't consider it a "god". It's not Zeus, Buddha, or Odin. It's not Jesus, Mohamed, or Ra. The Road isn't omnipotent and merciful, and it won't hand down wisdom from on-high, but it does hold the key, for me anyway, to everything I've ever wanted, and all the knowledge and history that ever was.

This is why I choose to be a Wanderer. This is why The Road is my home.

The Road is the path, the price, and the reward.

This book has been a labor of love. There are so many questions and considerations that you will face when deciding to, and eventually starting the process, of living as a Nomad, everything from what vehicle to choose to how many coats and shoes to bring along.

Make no mistake, the path you are about to travel isn't easy. You will learn things you never thought you would know, and do things that you never thought you could do. You'll make tough choices, and you'll make mistakes. - You'll make *a lot* of mistakes... Our goal is to give you some inspiration, and at least a few ideas about how it can all be accomplished, and hopefully help you avoid many of those time-consuming and costly errors.

Thinking back to the early stages of our personal adventure, we did what everyone does. We spent a lot of time daydreaming about what it would be like.

We imagined camping on the beach, boondocking in beautiful places, driving down the coast, climbing mountains, visiting national parks, posting breathtaking pictures, and getting thousands of followers.

Well, to be honest, social media wasn't really a thing back then, but we certainly did dream of someday writing a book. Of course, we were probably thinking more of a book detailing our amazing adventures, or perhaps a photo-log, maybe even a series of reviews or secret destinations (which we may still do). The book you are currently reading just seemed more appropriate at the time.

It is an understatement to say that there is a ton of information that needs to be gleaned, just for the process Building out a Bus or Van, or even those of you planning on traveling in a factory RV. You're going to spend a lot of time researching Solar, Water, Insulation, Appliances, to say nothing of Insurance, Budgeting, and figuring out how to make a living on The Road.

The gathering and documenting of that information can be daunting and confusing, to say the least, which is what led us to eventually start the website SkoolieSupply.com, essentially, our online guide to building and budgeting for your Nomadic Lifestyle. It contains in-depth tutorials on all of the topics above, and a great deal of additional information and ideas that will prove useful as well as educational. We've got something for everyone, Skoolies, Vanlife, Shuttles, Box Vans, even Truck Bed Campers. If you're not already familiar with the site, go check it out, and save yourself a lot of head-shaking and face-palming.

There is so much more, however, that needs to be considered, not just the nuts and bolts, but, perhaps, the "meat and potatoes" of actually living on The Road. The day-to-day activities and those small functionalities of a life in motion are easy to overlook, especially at first.

There's a lot of things that aren't exactly top of mind when you're trying to decide between a Shuttle, Van, RV, or Skoolie, or while you're combing the internet at 3am for the exact Build Space of a 2019 Sprinter (which is on our website, by the way).

The most common mistake that prospective Nomads make is attempting to do a Build that meets their current lifestyle, versus considering how they will adapt their current lifestyle to life on The Road.

We're going to repeat that, because it bears repeating.

The most common mistake that prospective Nomads make is attempting to do a Build that meets their current lifestyle, versus considering how they will adapt their current lifestyle to life on The Road.

Seriously. Read that again.

It's easy to fall in love with the high-end Van and Skoolie tours on the internet. - It's like Lifestyles of the Rich and Famous for Nomads. There is no shortage of online solar calculators, or folks willing to sell you a full-on reverse osmosis water setup, virgin cedar trim boards, or any manner of things that will make you look trendy and kitschy at the same time, but is all of that necessary?

Your life is going to change (whether you like it or not), and you should consider those changes before spending $20 grand on a Solar Setup that you don't really need, or building a professional kitchen that you will almost never use.

You may live on frozen pizza right now, or you might make huge vats of chili that could feed an army. Perhaps you just love to bake every day, or can't dream of living without your barista-quality, fast-heating, espresso machine, with the optional drip-tray.

There's nothing wrong with any of those choices, they just might not be the best ones you can make when cooking out of a Van. (We actually considered a Chapter/Commandment for this book called "Don't Eat Yourself Out Of House And Road", but it eventually fell under the heading "Mind Your Resources".)

The Meat and Potatoes of living on The Road are really about what you <u>have</u>, what you <u>want</u>, and what you actually <u>need</u> (more on this later).

We want to share some hard learned lessons and surprises that we've faced over the years, as well as some subtle life-changes that will make a huge difference, and can be incredibly beneficial in your Nomadic Journey.

Some of the ideas we present may seem odd , counter-intuitive, or even downright frightening at first. You may think some of these are unnecessary, or even ridiculous, but, as we said before, *"The Road has no memory and no mercy. It holds no obligations, and makes no promises."* You are on your own out here. Being successful on The Road isn't impossible. It isn't even particularly difficult, but it is certainly not a given. Making it <u>to</u> The Road is not the same as making it <u>on</u> The Road.

Think before you Build and keep learning as you go.

We've packaged these as "10 Commandments" but you can think of them as sound advice from some Old Skool Nomads, if it makes it easier to digest.

You will notice that we capitalize the phrase "The Road" as well as words like "Bus, "Van", "RV", "Shuttle", Skoolie", etc., and that we tend to use those words interchangeably. We do so because they are proper nouns, and because, as of now, they all refer to <u>Your New Home</u>.

The Bus/Van/Rig as well as The Road is where you are going to live. This book is designed to help you get there.

Commandment # 1

You Are Never in a Hurry

"Slower Traffic Keep Right."

As you've likely guessed, we aren't just talking about how fast you drive, walk, or power through that triple venti soy latte, but an overall attitude of living at a slower pace.

Living Nomadically means learning to appreciate your time in the slow lane, both literally and metaphorically, as the rest of the world zips by you on the left.

It also means taking your time as you plan and construct your Build. We'll discuss, later in the book, that it will never be "perfect", and you shouldn't be afraid to make mistakes, but taking the time to plan it all out, and think about what comes after whatever your working on at the moment, can greatly minimize the effects that those mistakes might have.

Taking it slow is the epitome of this lifestyle. In fact, it is the very point of this lifestyle. So few people allow themselves the privilege of doing so. Frighteningly few can even wrap their brains around the idea of stopping to smell the roses or taking in the view.

We Are those people.

Life is about giving yourself those extra moments to experience and appreciate your world and take in your surroundings.

As a child, you would have thought nothing of spending twenty minutes casually staring at a group of ants crossing the sidewalk on a warm summer day. It was pretty. It was interesting. It inspired you and piqued your curiosity.

As an adult, we've been told that we must run rip-shod over that sidewalk and ignore that moment, for fear that we'll be a minute late or the last in line. We have to keep pushing and moving forward. We're not allowed, even a few seconds, to appreciate the beauty of nature or the fascination of life.

That's bullshit.

The beauty is the point. You get one chance to experience this life. You owe it to yourself to take it all in, to feel every heartbeat and taste every breath. Take a walk in the park. Lie in the grass and stare at the sky. The moment you start allowing yourself the time to appreciate and celebrate every day is the moment you truly start to live.

If you're impatient behind the wheel, or if you're just, in general, a rush-rush, go-go kind of person, this lifestyle will probably be a very difficult adjustment for you. Make no mistake, however, it is probably the very adjustment that you need.

In a society is built for speed, we live and die by metrics. Shaving a few seconds off of mundane office tasks and production jobs has become a multi-billion dollar industry. Corporations have decided that it's acceptable to spend ridiculous amounts of money in hopes that they can convince people to put their stapler on the opposite side of their desk in order to save a millisecond of company time. Whether you call it "Best Practices", or "Lean Manufacturing" it all boils down to the fact that *Shaving a few seconds* is more important to your employer than your quality of life (or, by point of fact, your life, itself). Your supervisor has been instructed to "maintain a sense of urgency at all times", as that will subconsciously make the employees work faster.

Everyone you know can make a better hamburger than McDonald's, but none of us will ever sell as many hamburgers as McDonald's, because McDonald's gets those hamburgers to you really, really fast. - Our world rewards speed over quality and time over taste.

We drive fast. We eat fast. We even rush through what little leisure time we do mange to get because we've become so acclimated to doing everything hyper-quick and hyper-efficiently. (Go! Go! Go! Move! Move! Move! Run! Run! Run!) Over the years we, as a society, have kept steadily increasing pressure on the accelerator and pushing just a little harder each day, until we have no time to take a breath, let alone allow ourselves the luxury of a leisurely pace. We've literally forgotten how to take it easy.

That's not only bullshit, it's unforgivable.

If you are reading this book, chances are you have at least a moderate interest in changing that dynamic. You are curious, at the very least, about what other options there may be vs. charging ever forward, ever faster, into the abyss that has become our breakneck, get-it-done-yesterday society.

You've come to the right place.

Living on The Road certainly presents it's own set of priorities and "best practices", of a sort, but to choose this way of life is to thumb your nose at the status-quo. From the day that you begin your Nomadic Journey, you are on the path to a slower pace and more pleasurable existence. From the moment you start dreaming of converting a Bus or a Van, or just packing up the RV and pulling out of the driveway for the last time, you're moving toward a better place. - Freedom is closer than you think.

Certain aspects of this life will require you to "find a lower gear" whether you like it or not. You're going to spend a lot of time, literally, in the slow lane. Sometimes it will seem as though it takes days to climb a hill, and it might sound like the Rig is going to explode before you get there. You'll be passed by every imaginable type of vehicle, many of whom will immediately dart back in front of you and hit their brakes once they're there.

Alternately, since you're driving what is obviously a tourist attraction, passing motorists will cement themselves in the lane beside you, maintaining the same pace, as they stare at your Rig, mouth agape, as if they've just seen a UFO. (I've even considered dressing up like Bigfoot or Elvis while driving my bus... What the hell. Go big or stay home, right?)

If you are a road-rager, or even if you just get a little stressed out by traffic, you basically have three choices. Either you adapt and come to grips with it, let someone else drive, or you give up Nomad life for an activity where swearing is more encouraged (golf, perhaps).

Even if you're the most laid-back person on the planet, you'll still feel overwhelmed, especially when you first start researching all it will take to get you on The Road. This is a huge undertaking. It's akin to climbing a mountain or graduating from college. It is a life-changing process that will will bring you confidence and shape your mind forever. Everything from what type of vehicle to get, what you're going to do for a living, where to buy insurance, and even what color the pillows on the couch will be can seem like a life and death decision.

Once you start your Build, everything takes twice as long as it should, and half the stuff just doesn't come together the way you imagined it. Things will break. Designs won't work. Things won't fit. Mechanical issues, setbacks, busted knuckles... and you'll spend hours and hours standing in the aisle at Home Depot, redesigning a project because they only carried one size of something (and only in white).

It's Okay. You got this. Take a breath. Take a step back. It was like that for all of us. (It really was!)

If you choose to Build out a Van, Skoolie, Shuttle, etc. It will likely be the most difficult thing you've ever done. Your friends and family will tell you that you're out of your mind. You'll get dozens of answers and different advice from everyone you know (let alone the internet).

It won't be easy, but it will be worth it!

The Best advice we can give you here is to keep your Build Simple, and understand that it doesn't have to be perfect. It just has to work. Most of what you're doing can, and likely will, be redone after you're on The Road.

Give yourself time to enjoy the accomplishment of building your own home and your own space. Allow yourself to make those mistakes, take everything with a grain of salt, and let the naysayers say Nay.

You Are Never In A Hurry.

While it may not be as romantic or poetic as the sound of rain on the roof (or watching ants) and it isn't as tangible as enjoying the process of doing your Build. There is more to not being in a hurry than lowering your blood pressure and improving your attitude.

Slowing down and enjoying life may be the <u>reason</u> we do this, but Patience and Planning are what enable us to <u>continue</u> living life as a Wanderer. - It is an absolute necessity that you take your time and think things through.

While not being in a hurry will become your permanent mindset, Patience and Planning will be the difference between a long and happy Nomadic life or a short-lived, expensive, and uncomfortable vacation (one of which you'll never share the pictures).

For example, while you're allowing yourself the time to enjoy the process of your Build, it is of the utmost importance that you *take* the time to ensure that your electrical, water, gas, and mechanical systems are done correctly and will function safely. They don't have to be huge or elaborate. They don't have to use boutique-style products, or cost tens of thousands of dollars, but they do need to be safe. Patience in setting up your Systems is paramount. Double, triple, and quadruple check everything, and ask the experts if you're not sure (real experts, not just random idiots on the internet).

Even after your Build, your self-contained Solar, Water, and Propane Systems will require upkeep, and they will occasionally break down. It's surprisingly difficult to get an electrician or plumber to make a Van-call, and even if they do, your Systems will likely be much different than anything they've ever seen.

Mechanical breakdowns and maintenance can also be regular and seemingly constant activities, and the local Car Quest might not always stock the parts you need. You'll spend a lot of time working on the Rig, and it too has to be right. Once again, plan out your maintenance, and be patient when things go wrong. It's important to note that you are driving <u>your house</u> everywhere you go. Everything you own and the ones you love the most are in that house, and their safety and well being is in your hands.

Doing repairs and maintenance takes as long as it takes. Take your time. Get it right.

You should develop a pre-trip check list for whenever you're going to move the Rig. It should include checking fluid levels and tire pressure, belts, brakes, horn, lights, drive train... but also stowing and storing your gear as well as making sure nothing is loose or obstructing your path. - Truck drivers are required to log 15 minutes before each trip to do this. It is best if you spend at least 30.

Truckers are also limited in how many hours they can work. Driving a large vehicle can be stressful and exhausting. You're constantly watching your mirrors and checking your gauges, and you are attentive to your surrounding at all times. It's <u>work</u>. We've met more than one person who went through all the time and effort of completing a Build, only to learn that they couldn't handle the stress of driving a big Bus.

Be Patient. Pull over and rest.

Taking the back roads is one of the best parts of this lifestyle, and it's the best way to new adventures and things you've never seen. You have to be certain, however, before you head down that narrow dirt road, that you are going to make it to the other end, and that you'll have room to turn around once you get there.

Many state parks and campgrounds will display height and length restrictions, but most country roads and secondary highways aren't going to tell you that there is an overpass with 9' 8" clearance, or a hairpin turn that will leave you high centered across two lanes.

You don't find out until you're already there.

There are mapping apps, specific to large vehicles, that you can download to your smartphone that can help with this. Use those, check a topographical map, google earth, or just ask a local, but take the time to plan the route before you go.

You're also going to start adhering to the suggested speed on curves. - I used to take those yellow diamond-shaped signs with numbers like 30, 45, or 15 as a challenge, but in a Bus, Class A RV, or top-heavy Shuttle or Van those numbers are juuust right. You might even want to take it a little slower still.

Driving your house around isn't like jumping in the Honda and heading to Taco Bell (It probably doesn't fit through the drive through anyway). A little Planning and Patience can be the difference between a carefree day on the beach and a $1000 tow bill.

Finding a place to park presents it's own challenges, not just for the size, but for your personal safety. Parking lots and rest areas are usually okay, but can still be questionable. You never know who is going to pull up and take the space beside you. Don't be afraid to drive off and take your time finding an alternative. - Even better, have a backup plan in mind before you hit The Road in the first place.

When deciding where to stay for the night, you also have to ask yourself things like: Am I legal to park here? Are the bathrooms clean? Will it be quiet enough to sleep? Is there somewhere to walk the dog? Am I going to get boxed in? Do I have an out? What's my escape plan? The more forethought you put into this the better (and safer) life will be.

Planning and Patience.

You also have to consider your resources. You've only got so much water, propane, and food on the Rig, and running out at the wrong time (there's never a right time) can be devastating.

We dedicate an entire chapter to this later, but in short, you'll want to learn exactly how much water you use in a day, how quickly you burn through propane, and what foodstuffs are more practical for this lifestyle. You'll take the time to know how much you have on hand at any given moment, and be certain you're fully stocked before you head out of town for some boondocking.

Living on The Road requires Patience and Planning in just about every aspect of life. You're constantly cleaning, dusting, sweeping and putting things away. The thing you are reaching for is, most likely, behind or under something else. You'll spend a lot of time explaining to people about the lifestyle, how the Bus works, and arguing with random strangers about "just what the hell you think you're doing, anyway...".

Perhaps most importantly, living in a small space means taking your time with the people you love. Spending 24/7 with someone in a Bus or a Van can be taxing. Disagreements will happen, but you can't just walk into the other room and slam the door. (In most cases you literally can't.) Just remember that you are in this together and you're on the same side. Be patient and work it out. Allow yourself the time to listen and really hear what the other person is saying. Don't be afraid to be wrong, and don't be afraid to compromise (even when the other person clearly is).

Pro Tip: No one "wins" an argument, and the relationship is more important than being <u>right</u>.

There are dozens of little extra steps that we take every day to enable us to live this Nomadic adventure. Everything from switching the inverter on and off to handling your own waste, and dozens of other time consuming chores that simply have to be done.

The beauty of it is: Now you have the time to do all of those things. You don't have to rush through a meal or dart out the door. You can relish the extra time it takes to make dinner or clean up the kitchen. Changing the oil can be a labor of love.

Take a break every day just to watch the sunset. Pull over and see the world's largest ball of twine. Marvel at the magic and beauty that the universe has to offer, and play with the kids or the dog for hours every day. When things go wrong (and they will) you have all the time in the world to make it right.

You have <u>all the time in the world</u>.

To have the time, and not take the time, would be the biggest sin.

Some of us are patient by birth, and some of us have patience thrust upon us.

"Slower traffic keep right."

You Are Never In A Hurry.

Commandment # 2

Less Stuff = Less Stress

"If I don't need it, I don't want it."

If your house burned down tomorrow, and you lost everything in the fire, it would be terrible... devastating... But ponder this for a moment: You would be be forced to replace everything you own... More correctly, you would get to replace everything you own with only what you need, and only what was necessary and vital for your lifestyle. - There would be no more clutter.

The responsibility of Grandma's china, Aunt Gertrude's wedding album, or that weird statue of a rabbit that belonged to some funny-smelling relative that you had to hug and kiss at Christmas would be Gone... along with the stress of who to leave it to, or the guilt of throwing it away. - You would be free of the junk that binds you.

We surround ourselves with "stuff". We live in a world where shiny things, trinkets, and fancy toys are considered signs of success. We buy objects that we know we will rarely (if ever) use, simply because the fact that we *have one* somehow makes us feel better, more confident, or "part of the club". We grew up being told that the size of your house and how many luxury items you owned were a direct measure of your value as a human being. So we buy more stuff, then buy a bigger house to store that stuff, then we pay for a mini-storage for all the stuff that we can't fit in the house.

We're a nation of hoarders, inundated by advertisements urgently imploring us to buy the latest and greatest gadgets or suffer dire consequences. We're bamboozled into going into debt to purchase things we don't need, only to buy the upgrade a year later.

We're convinced that warehouse clubs save us money by allowing us to "buy in bulk" when, in fact, the per item cost is often quite similar to what you would find locally. It just comes in larger boxes. - More than you need, for more than you needed to spend.

We're so blinded by the addiction of consumerism and acquisition that we don't even do the math. "Bigger is better." and "Whoever dies with the most toys wins!"

Be honest. Someone you know (possibly you) has a large part of their garage or even an entire room in their house dedicated to storing items purchased at Costco or Sam's Club.

Open any cupboard and you will find an over sized package of _____? (Boxed Soup? Oatmeal? A case of canned hominy?) How long has that 10 pound bag of rice been sitting there gathering dust? Do you even remember when you bought it?

Without even thinking about it, we sit on products for months, if not years at a time because it was "a good deal". The concept of "Bulk" over-rides common sense, just like speed over-rides quality in our society.

In reality, everything you own has a price, and that price is much larger than what you charged up on your credit card as you walked through the checkout aisle. Sure, there's the monetary value and the sticker on the shelf, but that's not the True Cost. - That's not the Total. - In addition to the original expense, consider the time, space, and maintenance that the item requires, as well as the energy it takes, and the stress it creates.

A friend of mine is fond of saying, "There's nothing more expensive than a free boat.", and he should know. He has 4 of them. They're taking up space in his yard or parked on the street in front of his house, each in varying states of disrepair.

He's put a lot of time and effort into "fixing them up", but he puts substantially more time and effort into moving them around so that he can mow the grass underneath, or so they don't get tagged and towed by the local cops. He's constantly cleaning mold and patching holes, and it seems like he's always working on the trailer that he bought to haul the boats to the lake (which he does, maybe twice a year).

The True Cost of these free boats is the time, space, and maintenance they require, as well as the stress of worrying that he'll have to pay yet another fine when one of his neighbors reports the eyesore or expired tags. - What little enjoyment he gets out of these boats pales in comparison to the stress he endures and energy he expends just to keep them around ("But they were free!!").

This is why I am fond of saying, "What's better than having a boat is having a <u>friend</u> with a boat." All I have to do is buy a case of beer and throw him some money for gas.

Less Stuff = Less Stress.

Have you ever tried to donate a large desk or entertainment center to Goodwill or Habitat for Humanity? In most cases they won't accept them.

A lot of people get offended by the refusal of these items, thinking that their generosity is being snubbed, or the charity is just plain ungrateful, but in fact, they have just determined that the cost, time, and maintenance that these objects require is greater than the profit that the store is likely to see from the sale.

Habitat sells a lot of furniture, and their vast experience tells them that they can move several $15 end tables out of the same space and time it takes to sell one $30 desk. The end tables have a better return, they sell more quickly, and require less maintenance. Thereby, they are a far better use of the space and provide more benefit to the charity.

They only have so much retail area to work with, and they have to optimize how they "spend" it. The longer an item sits, the more often it needs to be moved, dusted, etc. to say nothing of the cost of disposal when the item doesn't sell at all. There is a real probability that the entertainment center you so graciously gave them shows up as a net-loss on the spreadsheet, once you factor in the True Cost of the item (maintenance, space, disposal...).

Your Rig is no different. You have to consider how you "spend" the space. The largest available School Buses have roughly 250 sq. ft. of "Build Area" (the available room behind the driver's seat once all the other seats have been removed). That's just a little bit larger than the size of two average bedrooms. A Van can be less than 70 square ft. (roughly that of a decent sized bathroom), and RVs often have several inefficient storage areas that are inconvenient to use in the first place.

One of the first awkward moments in planning for Nomadic Life is when you start considering how you're going to cram everything you own into a School Bus, Van, or RV.

Spoiler Alert: <u>You're Not</u>.

It's this early reflection point that often leads to the first disagreements between future Nomadic Couples and sends them off on a quest for a 40 foot pusher, or the biggest Class-A RV they can find.

This is where that "Costco-Thinking" gets us into trouble. The problem isn't that the space is too small. The problem is that you have too much stuff already (without even getting into all the things you currently have in your online shopping cart to buy for the Bus). Over sized coffee cans and giant bags of tortilla chips simply have no place in a Skoolie or a Van.

There's no shame or judgment here. We understand the cold comfort that comes from stocking up and having 200 double-rolls of toilet paper on hand. It's in our DNA. It's been pounded into our heads since childhood, and in your sticks and bricks house you've got plenty of room for it, so why not?

You can certainly build a walk-in pantry into a Skoolie, and some RVs even come with something similar from the factory, but consider the True Cost of the items. Consider how you could better "spend" that space. Instead of the comfort that you get from having 36 cans of green beans and 20 over-sized rolls of paper towels (hoarding) what else could you have? Extra coats and rain gear? Toy storage for the kids? A separate workspace? Emergency tools? A Generator? - Eliminating half of those rolls of Kirkland Signature Bath Tissue could mean 4 more sets of bed linen, or 400 amp hours of battery.

More to the point: How much time and effort are you willing to spend shuffling things around to get to what you actually want?

For each item in that shopping cart, and everything that you want to take with you, think about how much you are willing to move it out of the way, multiple times, every single day, to get to something else. - I call this "The Parable of Prepositions". It's the reality of living in a small space. Everything you own will be *on top of, under, in front of, or behind,* something else you brought along... Always.

No mater how good you are at Tetris, a year or two from now, you'll think about this idea and chuckle (as you're moving 2 Totes and a box of kitchenware to get to the pack of saltine crackers behind them).

Loading your Rig to gills with a month's worth of gear that won't be readily available where you're going is the exception and not the rule. For just about all of your Nomadic Journey it will make more sense to carry a few weeks worth of supplies, and restock more regularly, in smaller volume. - Have less. Move less. Use less.

Getting away from the "hoarding" mindset that we've been driven into our entire lives is not only healthy, but <u>necessary</u> for life on The Road. By DE-prioritizing things you don't need or use on an immediate basis, you are able to include more things that are useful, and even some that are just there for fun. ("Fun", by the way, is much more important than 2 extra cases of pork and beans.)

The good news is that, once you've been out here for a while, you will realize that the small space of your Rig (two bedrooms or a bathroom) is more than you actually need, and you'll feel <u>so</u> much better, not having to deal with all the random junk that you used to just "keep around" for (what you will learn) was no good reason at all.

It really does come down to what you have, what you want, and what you actually need.

Less Stuff = Less Stress.

So... In addition to closing the door on your "Costco Room" and never looking back, what trinkets and shiny things do you leave behind? What about those family heirlooms that you've been sitting on for years? How about the stuff in that black hole up in the attic? Which items that "We can't possibly get rid of..." do we haul off to the dump or force upon an unsuspecting sibling or cousin? What is important and essential? When is the last time you even looked though that box in the closet, or your high school yearbook? If you can't remember the last time you touched it, is it something you need at all?

Even without our addiction to *stuff*, these thoughts can be stress-inducing and emotionally charged. It's one thing to stop packing three weeks worth of food for a one week trip, it's another entirely to start getting rid of your old keepsakes and boxes full of memories.

Each one of us has a certain type of item that is more challenging to give up than others. Letting go of clothes, for example is very difficult for some, or shoes, perhaps. For many it's food, kitchen gadgets, books, or even pet toys.

For me it was tools. I've always had a full workshop holding a lifetime's accumulation of tools that I simply couldn't imagine parting with, but I had to ask myself, "Am I ever going to pour concrete again?" "Do I need a drill press and a wood lathe on The Road?" "How many screwdrivers can I use at one time?" - By means of confession, I gave most of the contents of my shop to a friend (the guy with the free boats from before) with the agreement that I could come by and use my old stuff any time I wanted. Even as such, I still have a few tools that I carry around, not because I regularly need them, but simply because they belonged to my grandfather. There's an emotional attachment that, to me, is worth the cost of the space these items require. I simply traded that space for less shoes and clothing items which, to me, were less valuable.

You may see it another way, and that's fine. Apart from Granddad's chisels and wooden mallet, however, there is nothing of mine on our Rig that I don't actually need for a specific and regular purpose.

Do that with your yearbooks and the kid's old Legos. Find someone who will hold them for you, or weigh their value against other items you may need, and swap them out. Just remember The Parable of Prepositions. How much are you willing to move them around, again and again, to get to what you need? (Granddad's chisels are on the bottom, by the way, because I know I'll almost never need them.)

Here is an experiment we suggest to help drive this home:

Go get a decent size cooler (or cooler-sized box). If you're a family of 4 or more, make it 2 coolers. - Open your fridge and take out only those items that you have used in the last week. Don't include soda, beer, leftovers, or anything that was part of a special meal or event, just the food-items that you typically and regularly use in a given week. You'll find that, unless you're quite the foodie, or a professional chef, these items probably don't even fill the cooler.

THAT is how much space you actually need for food.

Seriously. We know that you're trying to fit a full-size refrigerator into your Build. We also know that you probably don't need one. - What would using a smaller fridge free up space for? How much energy would it save? (More on this later.)

If you're a big beer drinker or soda-junkie, go ahead and call that a separate cooler (double the space), but for whatever is left in the fridge, reach in and grab the one thing in there that you can't live without. It could be a favorite salad dressing, a block of cheese, whatever's in the tinfoil in the back... When the cooler is full, you can only add something else by removing an item already in there. If you get stuck, consider alternative packaging. Does it have to be a full gallon of milk? Can you use the smaller size coffee creamer? Also consider items that don't actually have to be refrigerated. A jar of pickles or olives, as well as most condiments, are fine to store at room temperature. Make a place for them outside of the fridge.

By the time most people have completed this little experiment, they're actually looking at the cooler and thinking that they could cut back even more. Example: Whenever possible, buy perishables the day that you're going to use them, and only in the quantity that you will eat in a single sitting. - Yes, we know it it less cost-effective than buying in bulk, but it is way more "space-effective".

The spirit of this exercise isn't to deprive you of anything, but to start really considering what you have, what you want, and what you actually <u>need</u>. Do you really need it, or is it simply "nice to have around". Do you need a five year old jar of chocolate sauce or maraschino cherries? (No.) How long is it going to take take you to work through that 64oz tub of Country Crock?

One last important step: Before you put it all back in, take a look at what's left in the fridge and toss out everything that you haven't used in the last year (most of which is probably expired anyway). Really! Do this!

Nope... There's still more in there... Keep going.

Now, load up the items that you had in the cooler and make a mental note of how much room you have left. Take a moment to stare at it. Dig in deep...

The refrigerator in your house is a great example of The Parable of Prepositions. All the things that you had to move to get to what you wanted, and all the things in the back that you forgot were even there are a lot like those family heirlooms that you're still feeling guilt about trying to live without. - You can, and you'll be a lot better off for doing so.

Less Stuff = Less Stress.

In addition to considering a smaller fridge, let's take a look at the rest of the kitchen.

How often do you actually bake? When's the last time you used that $300 mixer? What about utensils? - Just like my screwdrivers, you can only use one whisk at a time, and how often do you actually separate eggs, anyway? Everything in your kitchen requires cleaning, maintenance and space, and you almost certainly need far less of it that you currently have.

Apart from making a holiday meal, when is the last time you used the full oven or more than one or two burners? Are there other ways to accomplish the same thing?

We use the microwave more than anything else, but when we do cook, we prepare almost everything with a portable cooktop and a Camp Chef Oven (also portable). We have a small hibachi or the campfire if we're looking for that smoky flavor, and when we simply have to bake bread or make cookies, the cost of checking into an Air B&B and using their oven is far less than the maintenance and space required to keep a 32" Amana Slide-In 4-Burner Oven in the Rig, to say nothing of the additional cost in Propane it would take to run the thing.

You may disagree, and by all means, go to the Cooking Page at skooliesupply.com and check out our selection of ovens. - If you absolutely love to bake, don't deprive yourself of the option. - If you don't however, or just don't do so on a regular basis, weigh the cost of this appliance vs. what else you could do with the space.

The same goes for the Air Fryer, Electric Griddle, Toaster Oven, Instant-Pot, Ninja Blender, and any other cooking gadgetry that you might have lying about on the counter or in back of the cupboard. - Prioritize it, consider the Parable of Prepositions, and don't forget the fuel or additional power it will require.

Sometimes it just makes more sense to check into that Air B&B once a month (which will also have a full shower, laundry, WiFi, and an outlet you can use as shore power to top off your House Batteries) and save the room in the Van for counter space or a convertible shower.

Now imagine doing the same experiment for your entire house... Because you're going to.

Whatever Rig you choose, you will be cherry-picking the entirety of your possessions, starting with the things you need to survive, the things you use daily, then weekly, and then prioritizing what's left to fit in whatever space remains. - What you have, what you want, and what you actually need.

The less stuff you take, the less you have to make room for, the less you have to move around, the less you have to fix, and the less weight, literally and figuratively, you will have on your back.

Everything that you <u>don't</u> take with you is one less responsibility. It's one less thing to clean, one less thing to repair, and one less thing to worry about falling off the shelf and landing on someone's toe. It's one less food item to spoil, one less dish to wash, and one less family heirloom to go missing somewhere in Montana.

It's one less Parable of Prepositions to store, shove, or move out of the way.

With every item you leave behind, you leave with it all of the power it holds over you. You are no longer a slave to needless items, nor imprisoned by things you can't give up. Every shiny trinket not taken is one step closer to freedom... and it gets easier. It gets a lot easier. <u>Life</u> gets easier.

If you still find this idea difficult to grasp, or you just want a backup plan in case you don't make it on The Road, then allow yourself to get a storage unit (or convince a friend who likes free boats to hang on to things for you).

On the day you are moving into your Rig, rent a U-Haul and park it next to your Bus, Van, or RV. Each thing you take from the house goes in one pile or the other. It goes in the Rig. It goes to the dump, or it goes in the U-Haul.

Drive that U-Haul, full of crap you don't really need but aren't quite ready to give up, down to storage and say "Goodbye for now." as you drop those things off. - I can assure you that, after about a year on The Road, you'll see that storage unit as a unnecessary expense and needless concern. The items that you left there will slowly fade from memory, and you will eventually give, or throw, those things away. - I speak from experience.

"If I don't need it, I don't want it."

Less Stuff = Less Stress

Commandment # 3

Mind Your Resources

"The old rules don't apply."

Living as a Nomad means changing your lifestyle. Not just where you work and sleep, but what you use and how much of it you consume. Some of the old habits and tendencies that we've developed over time are simply not applicable, if not outright counter-productive when living on The Road.

It is easy to become complacent when you live in a sticks and bricks house. You flip a switch and the light comes on. You hit the tap and water comes out. The shower is warm, the toilet flushes, and if we remember to take the cans out to the curb, a nice man in a noisy truck will haul away our garbage (never to be seen again). TV, Internet, and Mail Service are readily available. You can even "run for the border" at midnight, and be home in 30 minutes with 4 crunchy tacos and a burrito supreme. - Food, water, electricity, gas, everything we regularly use is just *there*. We don't even think about it.

This, of course, wasn't always the case. Until very recently, historically speaking, people essentially "lived off the land" in one way or another. You were directly responsible for your basic human needs.

As mentioned earlier, our hunter-gatherer ancestors followed animal migration trails. They did this not only for convenience, but out of necessity to survive. If they didn't hunt game, or gather berries and roots at the right time or season, they simply didn't eat. - There was no border to run to...

They didn't have department stores, an electric company, or a gas bill to pay. If they wanted to cook or be warm, they found firewood. They made their clothes out of animal hide and drank their water from the closest stream.

Our species survived this way for most of human history. For tens of thousands of years we hunted, found, or made what we needed and left the rest.

We are all direct descendants of Nomads and Wanderers.

Eventually, we learned to domesticate animals, planted crops, and became "subsistence farmers". If you could grow enough food and livestock to feed your family you were considered a great success. If you didn't... you were dead.

For the last thousand years or so, this is the way most of the human race survived.

Even at the turn of the 20th Century, power and gas were still a novelty. To have such luxuries at a private residence was considered opulence, an indulgence to be bragged about and shamelessly shown off by the ultra-rich. It was the exception, not the norm. Many houses in rural America weren't even hooked up to the grid until as late as the 1940s. - If they wanted to see at night they lit a lamp or used a candle. They cooked on wood stoves, and used a block of ice in a box to keep their food from spoiling.

We have evolved, over time, from a society of hunter-gatherers, directly responsible for our basic human needs to that of shift workers and consumers, exchanging our time for currency, and subsequently exchanging that currency for goods, services, and utilities.

In a sense, choosing the Nomadic Lifestyle and living on The Road means taking a step back and revisiting the spirit of our ancient ancestors. - No, you're not going to be tracking a woolly mammoth across the tundra with a spear, and you'll probably be better dressed than a 10th century Mongolian Sheep Herder, but you will be directly responsible for your basic human needs.

Everything that comes in or goes out of your Rig, the power you use, the food you eat, water you drink, the trash you create, and yes, the human waste, are all things that you will bear ultimate responsibility for managing and maintaining. Unlike living in sticks and bricks, where these resources are seemingly infinite, always available, and magically whisked away when discarded or no longer useful, a great deal of your life on The Road will center around acquiring, conserving, and disposing of the things you need to survive.

Mind Your Resources.

Growing up with infinite electricity, water, and gas has given us all a false sense of reality. One could even say we're spoiled.

If you're going to be successful on The Road, you're going to need to make some adjustments in how you use these resources.

There are some subtle changes you can make, and simple ways to conserve, without dramatically changing your lifestyle, that could potentially save you thousands of dollars on your Build, and hundreds of dollars per month moving forward in your life of wandering bliss.

They are also, very often, the difference between "making it" on The Road or heading back home, tail tucked, after a few difficult months.

Let's start with **Electricity:**

The idea of Solar Power often brings the most trepidation to those newly embarking on their Nomadic Journey. Most haven't ever used it. It's hard to understand, and there is no shortage of people on social media (and companies that are trying to sell you solar components) who will tell you that you *must* go with a huge Solar setup full of silver-stranded wiring, thousand dollar Batteries, and all manner of gizmos you've never heard of .

Alright, so, you've used one of those online solar calculators and figured out that you need 10,000 megawatts of panels and 20,000 amp hours of battery to run your Bus just like your apartment... Yeah... Okay... Cool... Great...

Well... That's not happening.

You (like it or not) are going to rethink your relationship with electricity, and you'll be better for doing so.

Building a gigantic Solar Array that will mimic your sticks and bricks house might be a viable option if you are willing to spend the money to try to make it happen. It is important to note, however, that even the most massive system won't be "infinite" (like what you currently have at home). You'll still find that you use power differently once you're on The Road. You will make certain compromises, whether your Solar Setup is 400 watts and 400 amp hours, or 30 panels and a big fat bank of batteries.

Many of these changes will be subtle enough that you'll hardly notice them. Most boil down to "be less wasteful" or "plan ahead".

If you consider adjusting your usage before you spend your entire Build Budget on things you might not fully understand, you can start out with a far less expensive Solar Setup, that will serve you just fine (and you can always scale up later if needed).

Trust us. You really can live this lifestyle without an upfront investment of tens of thousands of dollars in electrical components (and this comes from people who sell panels and batteries on our website).

90% of the time a Small Solar Setup is more than enough for your everyday usage. It will charge electronics, run lights, TV, and even a microwave without any problem. It's only when you use higher wattage items like an air conditioner, power tools, or anything with an element (see below) that you might over-clock the system. - Fortunately, there are easier and much less expensive ways to manage that remaining 10% of the time than "going big" on Solar.

Don't get us wrong, if you feel the need to build a Power Bank that could run a music festival, by all means, do it, but don't think that you *have* to, and don't let the cost and confusion of trying to put it all together be the deal-breaker that keeps you from pursuing this lifestyle. - It is simply not necessary.

Here are 6 simple steps that can make a smaller Solar Setup work for you:

1. Get a Generator: - "How many solar panels does it take to run an air conditioner?" - This commonly asked question is just plain wrong on many levels. - First of all, Solar Panels don't "run" anything, all they do is charge Batteries. The Batteries are hooked up to an Inverter that turns their DC Power into AC Power and effectively "runs" your 120v appliances (which includes the air conditioner). Additionally, it's not the number of Solar Panels, but wattage and battery capacity that matters, and the answer to "How much wattage and Battery does it take?" is *A Lot*!

You don't have to run all of your power-hungry 120v appliances from the batteries, however. You can also run them from from a generator. In fact, a generator plugged into the shore Power Connection on your Rig, will bypass the Batteries entirely, and with the addition of a Smart Charger can help top-off the battery charge as it works.

Sure, you'll have to carry around fuel to run the Genny, but they are surprisingly efficient. Rather than buying and hooking up twice the batteries and panels up front, you can get a Generator at a fraction of the cost (less than that of a single battery or a few panels). - See the Solar Page at skooliesupply.com

Think of it like using a wood stove to heat your home in the dead of winter. You don't need it year-round, but in December and January you might burn a cord of wood to keep from running your furnace constantly and paying exorbitant fees to the gas company.

Use a Generator, in the same way, for any high wattage needs, and then shut it off when you're done. If your goal is to run air conditioning, it will literally take years to offset what it would cost to build a Solar Setup big enough to run A/C by itself, and you're probably only going to use it a few months out of the year.

In truth, you will find that you use high-wattage appliances less and less, the longer that you are on The Road (including the A/C). You simply don't need them,

2. Use a DC-DC Charger to harness Engine Power: - A DC-DC Charger will allow you to charge your House Batteries from the engine alternator while driving. It's foolish not to take advantage of this power source that is already available in your Rig. It can do the work of several Solar Panels, especially on a long drive, or on a cloudy day. - How effective it is will depend on how much you drive, the size of your Battery Bank, and the capacity of your alternator, but this is a must have in any setup. - Even if you don't drive that much, it is a very effective backup system for when your charge runs low.

It is important to note that a DC-DC Charger only charges the batteries. It doesn't bypass them like the Generator does, as described above. Don't try to idle the engine to run your A/C. This can damage the alternator and potentially the Charging Unit as well.

3. Choose 12 volts over 120 volts whenever possible: - 120v uses 10 times the Battery capacity compared to 12v (not including the 15% - 20% loss during conversion), so when designing your Build always think 12v and LED lighting. Choose 12v options for as many of your creature comforts as possible. All of your electronics, including most laptops, can be charged from 12v dc. Televisions, projectors, all of your lighting, and many small appliances are available in a 12v version.

In our current Build, the Microwave, Water Heater, and an Induction Cooktop are the only 120v appliances that we regularly use. Sure, we have a crock pot and a convection oven, but we only dust those off when connected to Shore power (or fire up the generator if we just can't live without a Red Baron Wood Style Frozen Pizza while we're out in the boonies). Everything else runs from the 12v System. Our laptops, refrigerator, TV, Projector, and even a curling iron, are all 12 volt or rechargeable from USB.

4. Eliminate anything with an "Element": - An Element is a device that converts electricity directly into heat. An electric clothes dryer, electric stove, space heater, and even things like a coffee maker, toaster, or hair dryer create heat by running a large amount of current through a coil. It's a quick way to warm things up, but it comes at a remarkable cost.

A hair dryer, for example, will typically draw about 1800 watts on the Hot Setting, but only 100w on Cool. Running the element that heats the air requires 18 times as much power. A small Microwave actually draws less power than a hair dryer on the high setting because it uses different technology to create heat.

There are several options to work around this: Pour-over coffee, air-dry clothes, change your hairstyle (or wear a hat), Induction Cooktops, and several other items that will meet your needs without draining the Batteries. - Again, if you must use these high wattage items from time to time, do so sparingly, or as noted above, wait until you have Shore Power, or consider firing up the Generator.

5. Turn the Inverter Off when not in use: - One of the biggest advantages of using a 12v fridge (as well as the items listed in the last 2 sections), is that it allows you to keep the Inverter in the Off Position most of the time. Even at rest, the Inverter draws power, typically around 20 watts. (As a point of reference, your cell phone probably uses 4-5 watts.) While relatively minimal, it does add up over time and can shorten the life of the Inverter. The ability to shut it off, and leave it off most of the time, will make a big difference.

If you must use a 120v fridge, look into getting a smaller Inverter, dedicated just to that, so you can shut off the main House Inverter whenever not using it for the Microwave or Induction Cooktop.

There are more days than not, where our Inverter is switched on for only the 2 minutes that it takes to warm up breakfast in the microwave and then sits in the off position for the rest of the day.

6. Work With The Sun: - Rethink when you use electricity. We're all used to plugging in the phone next to the bed at night and leaving the TV on until morning. We eat a light breakfast and lunch only to "pig out" after dark.

Changing these habits is very simple, but will have a huge effect on overall Power Consumption. - Make sure your laptop, phones, Car Vac, and any other chargeable items are plugged in during the day. Even if you unplug it to use it, plug it right back in when you're finished. If you use electricity to cook, have your main meal in the afternoon and just a light snack before bedtime. Try to use heavy appliances while the sun's out, giving your Solar Panels the maximum time to collect sunlight and charge your Batteries.

You obviously don't have to follow all of these suggestions to the letter. As with Space and the Parable of Prepositions in the previous chapter, you choose your battles and priorities. If you absolutely have to use a 120v curling iron and don't have room for a generator, then bite the bullet and spend $1000 more on batteries (or plug it in in the bathroom of the truck stop).

These are just a few of the things that we have found to be useful. Get creative and come up with ideas of your own.

Rethink your relationship with electricity.

Mind Your Resources.

Please note, much more information regarding Solar Power and low-wattage appliances can be found at skooliesupply.com.

Now we'll open up the menu and talk about **Food:**

This is an area in which a lot of people are surprised. "What? I already understand food. I'm just going to keep eating the way I already eat, right?" Right!?!?

Well, you can, but it is important to consider what you eat and what it takes to store the things you eat, as well as cook your meals. As noted before, the bigger your fridge, the bigger the power consumption, and the bigger Solar Setup you will need to run it. You may opt to cook with electric, propane, or a combination of the two, but either way, these are finite resources you are using. This is another area in which the more you can minimize your consumption, the more likely you are to be successful on The Road.

This is a great opportunity to re-examine your diet and look for ways to lower your "food footprint", not just for power use and the environment, but your own personal well-being. In fact, one of the most impactful things you can do to conserve power is something your doctor has probably been advising you to do for years: Eat more fruits, raw vegetables, and leafy greens. Not only are you using minimal resources, but you're improving your health in the process. Raw nuts (almonds, walnuts, pistachio, peanuts, etc.) sprouts, and seeds, are also a win/win. Dried fruit and meats, or even certain prepackaged items like oatmeal, granola, raisins, and some protein bars are good choices.

If you read the label first, canned seafood such as tuna, salmon, and sardines can be surprisingly healthy and take zero energy to store or prepare. Make yourself a sandwich with quality bread and cold cuts, or go with the classic PBJ. A good breakfast cereal can be simple and versatile. Eat it right out of the box, mix with nuts and seeds, or have it with your preferred type of milk. There are even several good food choices in crackers and snacks. - Replace one or two meals a day with the items above and you'll be doing your body and your Battery Bank a favor.

How about some things that only take a few minutes to cook? Personally, I'm a huge fan of canned soups and vegetables, and I also go through a lot of Minute Rice. Granted, these items aren't likely to show up in the next Guide to Healthy Eating, and some of them might not do much for your Weight-Watcher's score, but if you keep an eye on what you buy, they can be both good *and* good for you. - Black beans and corn mixed with tomato or a little salsa is a great dip, side dish, or even a main course (it even makes a great burrito). There are relatively healthy options in baked beans, boxed dinners, and even canned meats that will cure your jones for comfort food and still keep you rolling.

It is also very helpful to multipurpose your food. Ham, for example, works for breakfast, lunch, dinner, and even a snack. You can carry fruit and veggies that are good to eat alone as well as for a salad or a smoothie. Bell Peppers can go in just about anything and are delicious raw, dipped in your favorite sauce, hummus, or salad dressing. Maximize your fridge space by carrying fewer items, that can be used in more dishes.

Obviously, we're not suggesting that the above items should constitute your entire diet, nor are we diminishing the wonder and beauty of cooking an elaborate meal in your Bus, Van, or RV. We're merely providing some examples of ways to be mindful of the resources available to you. When you have limitless electricity and gas you don't think twice about leaving the stew on simmer all day.

You can still do that in a Skoolie, but you need to be sure you have ample power/propane to do it before you start.

You should also be realistic about how much you are actually going to cook in the Rig. Lots of people say things like, "Oh, I'm going to cook so much more once we're on The Road. We need a big kitchen and all the gadgets...", but very few people really do this.

Actually, let's correct that: What we mean is that NO One really does this... Literally, like one in 100 people actually cook More once they're on The Road than they did before. - As we suggested with your Solar Setup, start with a minimal kitchen and scale up if necessary.

Be honest with yourself. Are you really going to BBQ that rack of ribs, or are you more likely to just find the local smokehouse? Are you the type to come home from the grocery store, put everything away, and then get right back in the car bound for Burger King?

How often do you Dine Out, and how often do you cook? Do you typically prepare a full meal or just heat something up? There are no wrong answers and no judgment in this. You just owe it to yourself to be truthful about it, and don't make yourself promises that you aren't going to keep. Don't spend a lot of money on kitchen appliances, propane, electricity, and groceries that you might never use.

Even if you are a bit of a kitchen hound, consider what you can make with the items mentioned above, or look into recipes that use low-impact (on power) ingredients. There are plenty of options.

One of my favorite parts of this lifestyle is finding those roadside diners and dive-bars that offer amazing food (often only known to the locals). It's difficult to put a price on appreciating the local flair and culture, and we enjoy dining out. We keep less spoilable food in the Rig, knowing that we are more likely to find a good seafood joint than we are to actually make that "Smoked Salmon Surprise" recipe that we saw on Pinterest.

Once again, this is an area where you will prioritize and pick your battles. As with most Americans, I grew up loving fast food. It was a staple of my diet well into adulthood (and long after it shouldn't have been).

I was helping a friend with a project the other day and he drove us through McDonald's for lunch.

I have to confess that I was a little excited, having been "off the stuff" (anything that comes out of a drive-through window) for several years now, but when we pulled up to pay, and the total was almost $40 for the two of us, I felt my enthusiasm immediately start to wain. Don't get me wrong, the fries were still sinfully good, and nothing tastes like nostalgia, but it reminded me of why I stopped going to TacoWendyMcCarl'sBurgerInTheBox right about the same time I started living on The Road. - It just isn't worth it. - By no means am I a health nut or anything, but since I am "Never In A Hurry", I have ample time to make better food choices.

More to the point, since I stick to a pretty strict budget, I choose dive-bars and roadside diners over fast food. - It wasn't even a conscious decision. I just stopped doing it (the Rig didn't fit through the drive-through anyway).

Whether for health or financial reasons, you will likely start prioritizing the types of establishments you frequent and saving your going-out-budget for the ones at the top of the list.

You may love food trucks or five-star restaurants. You might only choose to dine out at famous tourist destinations like Cafe' Du Monde or Gino's East Pizza. You might even think of my my dive bars the same way I consider fast food, and that's fine. - If you have unlimited funds and want to go out to eat every day, by all means, do it! If you're on a budget, weigh your options and prioritize. If you *never* dine out, you love to cook, or you're the one in 100 we mentioned before, go with a full propane oven, multiple burners, and get a bigger fridge (and dedicate an Inverter to run it).

Just be honest with yourself upfront about how you're going to use food, and consider what it costs to keep it around.

Mind Your Resources.

Alright. Let's hit the showers and talk about **Water:**

Water is the most vital (and endangered) resource that we humans rely upon, while also being one of the most taken for granted. Many new Nomads are surprised at how much time and effort they spend looking for and acquiring Water, as well as how quickly their Fresh Tank is emptied. Campgrounds, National, State, and County parks, Rest Areas, and most Truck Stops are good places to fill up, but it may not always be available or accessible.

Pro Tip: - It is better to "top off" your water tank whenever you happen to find potable water, rather than waiting for it to be nearly empty before you start looking.

It will be a large part of your regular routine. You'll want to look at your water tank every day, monitor the level, and keep a mental inventory of how fast is being depleted. You can think of it as a "water budget". - "We started with 100 gallons and we use 10 gallons per day, so a full tank is 10 days worth of water." ~or~ "We have 40 gallons and we use 2 gallons per day, so we have 20 days."

There are many variables in the amount of water you will use. It will depend on what type of toilet you have, how much coffee you make, your pets, showering habits, cooking and cleaning, and yes, how much water you actually drink. It will also depend on how you construct your Water System and how you actually use it.

Your Water Setup might be as simple as a jug that you pour into the sink and drain into a bucket, or a full on Plumbing System with a City Water Inlet, pump, and filters, but regardless of how complex is is, the following tips will help you conserve this vital resource and stretch your water budget to the fullest.

1. Shower responsibly: - There are few pleasures in life better than a long, luxurious, hot shower. We love to stand under the water and wash the day away. We use it to relax, to wake up, to clear our thoughts, even to make important life decisions. For many of us, the shower is the only time we get to ourselves, so we take advantage of it.

Sadly, unless you have an infinite source of water, once you're living on The Road, a long luxurious shower will be a thing of the past. Don't get me wrong, it can be done, and there are alternatives and options that will allow you to indulge from time to time, but for the most part, a "Navy Shower" will be the order of the day.

The term "Navy Shower" typically means: Turn the water on, Get wet. Turn the water off. Lather up. Turn it back on and rinse off. The water should be on for about a minute each time, and should get you clean with only 2-3 gallons used. - You can cut that by another gallon or so if you wash your hair in the sink (substantially more, if you wash your hair in the sink of a truck stop or a rest area).

The reality of living on The Road is that you will probably have a sponge bath (or use wipes) much more often than you will actually take a shower. Most Nomads do some version of a sponge bath every day, with a Navy Shower only happening once or twice a week.

A gym membership is the easiest way to expand your options. For $10 - $20 a month you can still have your long luxurious (or even daily) shower, as long as there's a Planet Fitness within driving distance. For Van-Dwellers or those in a small RV this is the way to go, but even if you have a full size shower in your Rig, it is money well spent.

You might also consider building a Recirculating Shower, with which you can shower as long as you like on a gallon or two of water. - See the "Unique Build Ideas" Tutorial at skooliesupply.com for an example.

2. Split the tap: - You need to decide, early in your Build process, if you are going to drink the water from your fresh tank or if you will only use that for cleaning and bathing. It may seem counter-intuitive, especially if you're building an elaborate Plumbing Setup, but there are some valid reasons to keep on buying bottled water for drinking and cooking.

First and foremost, you can't always trust the source. Let's be honest, that's why we buy bottles and jugs of water now, right? - If you don't trust the water at your house or office, why would you trust it from an outdoor spigot at a Flying-J on I-40?

There is no shortage of YouTube videos showing elaborate filtration and reverse osmosis water systems, and there's no reason you can't do that, but take a minute to weigh the cost of that setup (including space, time and additional filters/chemicals) vs a buck or so a gallon for drinking water (50 cents or less if you refill at WinCo/Walmart/etc.).

Yes, the expense of buying drinking water will add up fast. Building the filtration system might cost less over time, especially for a large family. Weigh this decision carefully, but for us, having done it both ways, the simplicity of a Plumbing Setup that requires (almost) no maintenance, coupled with the peace of mind that our filters aren't bad and our water isn't tainted, as well as substantially less time spent filling the Fresh Tank, make it well worth the cost of keeping jugs of drinking water on hand.

3.Don't let it stream: - Have you ever met someone who was so weird about water conservation that they brushed their teeth or shaved using a cup? - Well, that's going to be you. - You'll also be basically doing the Navy Shower thing with your hands: You'll turn the water on for a second and get your hands wet. Then turn it off, lather up, and wash vigorously. Then use a quick blast of water to rinse.

Brushing your teeth or shaving with a cup, as well as Navy Shower Hand Washing, is something we should have all been doing since we were kids. To let the water stream on full blast while we casually scrub our choppers, scrape the whiskers, or pump that foamy soap on our hands is incredibly wasteful.

Two minutes to brush your teeth could be 4 gallons of water, and five minute shave, as much as 9 or 10. - This goes for the ladies as well. A small bowl of water is plenty to shave your legs. Do it before you take a shower, not while you're in the shower.

4.Change the Aerator: - "Standard" water flow for a kitchen sink has been 2.2 gallons per minute since roughly the dawn of human civilization (or the 1940s, whichever came first). The aerator is the device on the end of the faucet that controls how much/how fast water comes out. They're usually easy to replace, pretty inexpensive, and available at any hardware store in sizes ranging from 2.2 all the way down to .5 gallons per minute.

Many newer faucet sets are non-user-serviceable, but typically come with a reduced flow already installed (read the box before you purchase). - This simple move can save 10-20 gallons a week, even for those who are already using water conservatively.

By the way, the same theory applies to your shower head. Shop around for one that is 1.5 gpm or less and make your Navy Showers even more efficient.

5. Coffee: - I recently quit drinking coffee, and I have to say, it's made a huge difference, not just in water consumption, but life, in general. - Of course, I'm not suggesting that level of insanity to anyone else. A friend of mine refers to coffee as "anti-murder-juice", and she has a point.

What you can do, however is be more realistic about how much you drink. I actually remember having heated discussions, in years gone by, about how "...making less than a full pot of coffee at a time is ridiculous!" While my arguments were compelling, it turns out I was dead wrong. In sticks and bricks, with unlimited resources, you can make a full pot and let it sizzle all day, only to dump 1/3 of it out the following morning, with little concern. Living on The Road, however, with limited water, you need to only make what you drink. You'll likely be making pour-over / cowboy coffee anyway, so make less of it at a time (you can start practicing this now) and get a feel for what you actually use. If you are a 1 or 2 cup of coffee per day person, a Keurig is fine (even though they're a little heavy on power), just be realistic.

6. Laundry: - Both for Water and Power (as well as Space), Laundry can be a huge drain. There are some efficient combo-washer-dryers, and there are some smaller, portable units that work quite well (check the Water Page at skooliesupply.com), but a load of laundry can easily burn 20 gallons of water, even with the more efficient models. Dryers are very heavy on power consumption.

Ultimately, given the upfront cost, space, and continued use of water and electricity, you'll probably find out that to pony up the quarters for the local Suds-O-Rama is more cost-effective than trying to wedge a full size W/D into your Build.

If you can do smaller loads, a manual unit, like the Wonder Wash, will do the trick. Use a salad spinner for an extra spin cycle, then hang on a line. We've also known folks that conveniently "visit a friend" right around laundry time and "....Oh do you mind if we toss in a load while we're here...". Some have been successful with a washboard in the creek, and I once knew a guy who used a toilet plunger and an old cooler. (Don't worry it had never been used to plunge anything else.)

Alternately, as we mentioned earlier in this book, spending a night an Air B&B is a viable option. Toss in a load of laundry, do some baking, use their WiFi to upload those videos you've been working on, drop an extension cord out to your Shore Power Connection and charge the batteries, and take that long luxurious shower while the clothes are in the dryer.

If you're dead set against laundromats, smaller loads are just impractical for your lifestyle, you're too classy to mooch off of friends and family, and you have the space, go ahead and buy a combo unit from our website. Just plan on staying at a park, hooked up to Shore Power/"City Water", every time you need to do a load.

As with everything else we're suggesting in this book, take it to whatever extent you feel comfortable. The more of these suggestions you follow, the better your chance of success on The Road, but even small changes can make a big difference.

Mind Your Resources.

Time to feel the burn and save a little **Propane:**

To be honest, the best way to save Propane is to not use it at all. To go "Propane Free" (or at least very minimal) in your Build, is quite doable, and a good idea for several reasons.

"LPG" (Liquid Propane Gas), is a hydrocarbon byproduct of fuel refining and Natural Gas Processing. It is not the same as Natural Gas ("LNG" or "CNG"). It is a common mistake to think that a CNG Bus runs on Propane and you can power your Propane Appliances from the same tank. - You can not. They are different fuels.

In addition to being a hydrocarbon, LPG is a finite resource, and prices can fluctuate greatly, not only with the cost of gasoline or crude oil, but also with the passing of the seasons. Demand goes up Summer and Winter, and it tends to drop in early Autumn.

Couple this with the time and expense of mounting an LP Tank outside the Rig, running lines, a Manifold, Pressure Regulator, etc., installing appropriate venting, and the possibility of a gas-leak caused by bouncing and shifting while driving down the highway, and it's easy to see why going "Propane Free" or "Propane Light" is becoming more and more popular.

Whether you do it for environmental concerns, the cost and effort of building an additional system, the fluctuating price of ongoing consumption, or just for safety, you can minimize or eliminate your use of Propane/LPG, by simply using different appliances.

Propane is most commonly used for heating the living space, cooking, and heating water. You can easily go "Propane Free" by using a Diesel Heater, Electric Water Heater, and eliminating that full-size oven. - We accomplish most of our cooking with an Induction Cooktop and a Microwave, both of which run on Electricity, which, even with our minimal Solar setup, is *essentially free*. We re-create the energy it takes to do our cooking, as well as what we use for the Electric Water Heater, as opposed to venting expensive gas into the atmosphere.

The Diesel Heater, of course, does require an Exhaust, and uses a consumable fuel, but the efficiency of these units is enough that the impact is minimal. - The best solution to minimize heat use is to "Follow the Weather". Try to stay where it's warm and don't ever turn the thing on in the first place.

Choose your battles: - Of course, there is probably going to be some Propane somewhere in your Rig. A Mr. Buddy Heater, and a Coleman Camp Stove as supplemental/backup appliances for heating and cooking are almost a necessity. They are portable and run on 1 pound bottles, thereby saving you the expense of installing a full Propane Setup and mounting tanks outside the Rig. - While you can "temporarily" transport up to 21 gallons, it's illegal, per DOT, to carry anything larger than 1 pound bottles inside a vehicle.

If you're a heavy baker, live on frozen pizza, or simply aren't ready to part with certain comforts of home, having a full oven might be a necessity. - If the oven is your only major Propane Appliance, consider running it from a 20 or 30 pound bottle (mounted outside) using a single line and onboard regulator. This will be much more simple than a full setup, and give you the option of "swapping out" the bottle or easily removing it to refuel.

If, however, you only bake "once in a great while", or if you just "want the option" of baking if it comes up, consider a portable camp oven like the Camp Chef or similar (skooliesupply.com Cooking Page). This, like the Buddy Heater and Camp Stove run on one pound bottles and can be tucked away when not in use.

To those just beginning the research for their Nomadic Journey, some of the ideas we've presented in this chapter may seem extreme, ridiculous, or even frightening. Anyone who has been on the Road for awhile, however, is probably thinking of a half-dozen things that we forgot to mention.

The simple reality is that you will adjust the way you use electricity, food, water, and fuel when living on The Road. Even if you spend the extra tens of thousands to try to mimic your sticks and bricks power, water, and fuel systems, you will come up short, and find yourself devising creative ways to stretch your budgets in each of those areas. It's a necessity.

The irony is that you will actually find life to be more enjoyable in so doing.

We human beings are actually not, genetically, a wasteful species. We've just become complacent over time. For most of our history, we took what needed, left the rest, and we moved on. - In the same way that having less stuff equals less stress, *using* less stuff is actually our natural state of being, and it brings a freedom all its own.

"The old rules don't apply."

Mind Your Resources.

Commandment # 4

Keep It Simple

"Everything works if you let it."

The above is a quote from a cult-classic movie called "Roadie", which stars the late vocalist, "Meatloaf", as a guy with the rare ability to fix nearly anything that's broken. He repeatedly gets coerced into working road-crew for various bands (Alice Cooper, Hank Williams Jr., and Blondie to name a few), and he uses some very creative methods of running sound and power for the shows.

One of my favorite films of all time, it also includes the quote, "Anything worth doin' is worth over-doin'.", which is, of course, completely contrary to the central theme of this book. We've spoken quite a bit, so far, about <u>not</u> over-doing or over-complicating things, both with your Build and with your life.

Learning to live free, as a successful Nomad, is about taking a step back, examining the different habits and unhealthy desires that we've developed, and undoing the programming behind them. We must be willing to unlearn what we have learned, and dismiss the false narratives that we've had drilled into us. We have to turn down the noise, let go of complications, and release ourselves from false aspirations and expectations.

Keep It Simple.

Sadly, this is not how our society likes to function. We often go out of our way to make things harder than they need to be. Our lives are convoluted, over-done, and complicated. We wear it as a badge of honor. It's an unwritten social "norm".

To respond with the phrase, "It's complicated." when asked about your life, work, or a relationship, is almost guaranteed a round of murmurous laughter. - They don't really know what they're laughing about, but do so out of nervous empathy, as if to say, "Oh. I feel ya! My life is complicated too! blah blah blah..." .

Why do we do this?

Where in our history did we decide that enjoying the simple things was to be frowned upon, and living an overly-complicated life was a sign of success?

We seek out complications and problems. We look for conflict, and go out of our way to make issues where none exist. We borrow trouble, and find excuses to argue about matters that are relatively inconsequential. So many of us are so entrenched in gossip and drama that we have entire TV networks dedicated to it. We are addicted to Reality TV, not because it's "real", but because they hyper-focus on the conflict and the complications. We don't really care if people survive in the woods, which one gets to marry the hot bachelor or bachelorette, or even who wins the karaoke contest. - We're addicted to the drama. The more conflict and complications, the better.

MTV and VH1 realized, early on, that more people would tune in to watch idiots yelling at each other than would tune in for the music that these channels were originally created to air. The majority of cable networks and streaming services now offer more shows that can be classified as "Reality TV" than all other programming combined. - It's over-done, contrived, a complete waste of time, and we just can't get enough.

Strewn among these highly-staged docudramas and faux-reality shows are an endless stream of advertisements. They use our inability to look away from the train wreck to sell us shiny trinkets and beer... and it works!

We've been told what we want since we were children. It's been beaten into us relentlessly by people trying to sell us everything from baby products to adult diapers. They show us what they think we want, so they can sell us what they say we want, and we buy it, hook, line, and sinker.

Convincing us that we need something today, that we didn't even know we wanted yesterday, is a billion dollar industry. They're so good at it that most of us have great difficulty in even distinguishing our wants from our needs (with no regard at all to what we already have). We need food, but do we need fast food? We need clothing, but does it have to be Abercrombie? Do we need a Corvette or do we really just need a Geo? - The church of the almighty advertisement permeates every aspect of our lives.

To be successful on The Road, in fact, to even make it <u>to</u> The Road, we need to to stop listening to those voices. We have to drop the drama and stop seeking out complications. We can't feel like we have to over-do, over-work, and over-spend, just to get the things that we've been told we want. We have to ignore the pro-athlete telling us to eat food that they've probably never tasted and focus on what we actually need to make this work.

Keep It Simple.

It is the same programming that makes us accept those unnecessary complications and buy the shiny trinkets that convinces us that everything in our Bus, Shuttle, or Van has to be absolutely perfect.

It doesn't.

It has to be comfortable and it has to function, but it should be about what you have, what you need, and not what someone else tells you that you want.

We've all seen the videos of Buses and Vans that look like a five-star hotel or luxury destination. They've got the newest top-end appliances, real hardwood floors, cedar ceilings, and beautiful rooftop decks. You see Skoolies and Vans with professional kitchens, custom paint jobs, live-edge... everything..., hammered copper sinks and tubs, full tile walk-in showers, and fabrics seemingly woven from the tears of angels. You see the custom shops that make teak-wood drawers, automatic sliding bed platforms, theater lighting, and full-on surround-sound stereo systems. They are plush, posh, and perfect from wall to wall.

Don't get us wrong. A lot of that stuff is really cool, and it is absolutely vital that you make your conversion feel like a home. It needs to be comfortable, and you should be happy with the space you've created. You want to be confident having company over, and feel good about showing it off. You worked hard to get here. You should be proud of the new skills you've learned, and all of the things you've accomplished along the way, but it is very important that you don't get in over your head, or try to over-do it, especially if this is your first Build.

Do the hammered copper or the live edge if that's what you really want. You can buff, sand, and seal those floors to your heart's content if that makes it feel like home, just don't feel like you <u>have to</u> do all of those things to be accepted by your fellow Travelers. In the real world, we are much more impressed by something converted, re-used, or re-purposed to function, than we are about $2000 fireplaces and teak-wood flooring.

"Everything works if you let it."

The people that you see on social media did what they did for the specific purpose of becoming "Influencers" and making a living through their images and exploits. Like it or not, they are our version (the Nomad version) of Reality TV.

If that is your goal, then by all means, go for it! If you want to be "Lifestyles of the Rich and Famous on Wheels", we're not going to stand in your way. - If, however, your goal, is to be successful living on The Road, you need to be a little more realistic about what you're doing, and think more like "Lifestyles of the Happy and Easy-going". Or, perhaps "Lifestyles of the We Didn't Blow Our Entire Life Savings on the Build".

Keep It Simple.

This applies not only to the materials you use, but to how you actually construct your conversion.

Those folks from the last chapter with the reverse osmosis water systems and huge solar setups will tell you that there is only one way to do it. They'll tell you that you have to replace all your windows, cover the seat holes with pennies, and spray-foam from top to bottom (among so many other things).

The Skoolie and Vanlife groups and forums are full of people who seemingly love to start arguments about someone else's suggestions. They'll do so to the point of calling names and ridiculing the other person, as if to suggest that entertaining any deviation from their own personal advice is akin to being in league with the devil.

Does any of that sound familiar?

The reality is that these people are either horribly insecure, or they're trying to sell you something (probably a little of both).

Yes, it's true. We at skooliesupply.com are also trying to sell you something, however, we are not going to tell you that you absolutely <u>must</u> to do things one specific way, or that you have to buy the biggest and most expensive items (even though we would make more money on it). We're not going to tell you that there is only one option, and anything else means utter failure. - Because it's simply not true. - Your Build is your business. There are multiple ways to accomplish just about everything, and while some choices might be better than others, it all really depends on your personal situation.

We could literally fill up this book with examples, but in addition to the above, here a just a few commonly touted "only one way to do it" suggestions that are often needlessly time consuming, confusing, and can be very expensive for those of you just getting started:

1. <u>Never use any products from company "x", or Only use products from company "y".</u> : - Okay, seriously, these are the Ford vs Chevy people. These guys will tell you that every Ford ever manufactured is better than every Chevy ever built (or vice versa). Is that true? - No. Of course it's not. - There's nothing wrong with brand loyalty (perhaps), but to *never* use Renogy (for example) just because some dude on the internet said not to, is just plain silly.

We've used their products in dozens of Builds and never had an issue. - This goes for Roof Vents, Water Tanks, and a dozen other items commonly found in a heated online debate. Even if they can argue that another company makes slightly better quality products, "Never" and "Only" are pretty strong words. - Do your research and figure out what works best for *your* Build, not what the loudest voice in the forum keeps repeating.

2. <u>You must use Spray Foam and a Vapor Barrier</u>: - Nope... You certainly <u>can</u> use Spray Foam, and in certain situations it's the best choice, but the extra work, costs, and potential off-gassing can end very poorly if you're not very careful. This is another area in which the guy from the previous paragraph kept typing in ALL CAPS, and so many people have just regurgitated what he said, that we assume it's common knowledge. There are, in fact, several options, and which one is best for you depends on your Rig, your budget, and your situation, not to mention your feelings about the environment.

In our opinion, Spray Foam is probably a second or third choice, and a Vapor Barrier is <u>almost always</u> a bad decision. Read the Insulation Tutorial at skooliesupply.com for more info, and then make up your own mind based on your own personal needs.

3. <u>You have to eliminate all Thermal Bridging</u>: - This one is actually based in scientific fact, and you should make a reasonable effort to mitigate as much of it as you can, but to attempt to eliminate <u>all</u> Thermal Bridging, everywhere, is a pretty tall order.

In the early stages of your Build, you'll be faced with the concept that the outside air temperature is conducted into the living area of the Rig via any contact made with the skin of the vehicle. In other words, if the wood floor you build inside actually touches the metal floor of the Van or Bus it defeats the insulation. - This is why you see so many posts about laying a plywood sub-floor directly over rigid foam insulation, without any cross-members or floor joists present. They're trying to prevent any part of the living area from touching the outside metal of the vehicle.

Do your best to minimize thermal bridging with Butyl Tape or even Construction Adhesive (use it to affix the joists), but you don't need to spend a month trying to isolate every single surface, and don't tear things out that you've already completed just to retrofit for this purpose.

There are dozens, if not hundreds of examples of these "only one way" arguments, from using stranded copper instead of solid wire, or Birch or Marine Plywood instead of CDX. Many, if not most, of these have a basis in fact, and might even be a "better idea" (within reason) but they are really not the life and death decisions that the ALL CAPS people suggest.

With the people shouting "only one way" from the rooftops, all of the contradictory information on the web, and so many absolutely beautiful Buses and Vans on video that must have cost more than the GNP of a third world country to complete, it is easy to get overwhelmed. It is easy to get sucked in with the shiny trinkets and feel like you have to conform to these very high expectations.

You Don't.

The harsh reality is that if you try to build a conversion that will compete with those "luxury resort" Busses and Vans that are internet-famous, while listening to all of the "only one way" arguments, you are almost certainly going to fail. Roughly One in Four people who start a Build never complete it, for this very reason. It's too difficult, they run out of money, or they just give up because they can't handle the stress of everything having to be "perfect".

Another reality, one that is surprisingly unspoken, is that nearly half of those that actually "make it" to The Road only live in their original Rig for six months to a year.

Some just fail. The lifestyle simply isn't for everyone. Perhaps they can't handle such a small space, can't live within their budget, or lose their income stream. Some just don't feel safe. It could be the driving or the sleeping in strange places, but lots of folks just never get their sea-legs, and they go back to sticks and bricks.

Even for those that do make it long-term on The Road, your first Build is likely not your Forever Home. There are exceptions to this rule, to be sure, but a typical Skoolie-Dweller only lives in their first Bus for 1-2 years at the most. Vanlifers tend to have a little more time due to less maintenance and fewer breakdowns (or at least more simplistic and readily available repair options), but even if your Rig is in great shape, there's a real possibility of out-growing your Build, or just realizing that you should have done things differently.

Yes. It's very common that your first Build tends to be "wrong". Three months into your Nomadic Journey you are wishing that you hadn't bothered with the walk-in shower and built an extra seating area instead. You don't really use that cool shoe-cubby, and you keep hitting your head on the spice rack (and the spices keep falling out anyway). The things you actually use day-to-day don't fit in the storage areas, and what the hell was I thinking with this color scheme?

The first time you do a Build, you have very little idea what you need. You're not really sure what to prioritize, or what you actually want, and no one can tell you, because it is different for each and every one of us.

You won't know what you truly <u>need</u> in the Rig until you have lived in it for a while. If you've spent all your money and permanently mounted everything in place, and that place turns out to be the wrong place... You're kinda out of luck.

Don't get discouraged. - If you are dedicated to being a Nomad, you will figure it out. There are no shortage of remodels on The Road and plenty of fellow travelers that are more than willing to help. Just bear in mind that the less you over-do it from the start, and the more you "let it work", the easier it will be to revamp, redesign, and rebuild once you're rolling.

The best idea is to make your first Build as minimal as possible, make sure that everything can be easily removed or replaced, and use less expensive or even temporary materials (you can still make it look good). If you spend less time and money trying to make it plush, posh, and perfect, not only will you substantially reduce your stress level, but you'll get on The Road sooner.

Then, just like remodeling a room in our house, you can upgrade, as needed, once you have a better idea of how to make everything work, and what things just plain don't.

Save the money you would have spent on reverse osmosis, and use it to add that hammered copper six months down the line, after you've figured out that you need to relocate the sink anyway.

The whole reason you are choosing life as a Wanderer is to keep things simple. Less drama, less stress, less pressure, in life as well as in your Build.

"Everything works if you let it."

Keep It Simple.

Commandment # 5

Maintain Thy Rig

"Keep it runnin' like a top!"

Early on in your Nomadic Journey you will start to have some very awkward conversations with family and friends. "You're going to live in a what?!" "You want to be homeless?!?" You'll also endure endless and ever-present "...van down by the river..." jokes.

We've even met multiple people who said that their mother cried, actually broke down in tears, when they spoke of it. "Did I fail you?!" "Move back in with us if you can't afford you're apartment!" "Whatever will I tell the ladies in the Bridge Club!?"

Okay, that last one I just made up, but I imagine that many parents are thinking something along those lines when their kids first present them with the idea of living Nomadically.

Why?

Because they just don't get it. To them, living in a Bus or a Van means those moldy, broken down hunks of rotting steel parked under the overpass and surrounded by junk. That mental image permeates their thoughts because they've seen it so many times on the news, usually accompanied by an interview with some neighborhood committee member demanding that the local mayor or sheriff *do something* about this scourge of derelict free-loaders.

They shout, brow furrowed and fist shaking in the air, "They're all dirty drug-users and we want them OUT!" "We don't care where they go, but not in OUR neighborhood!"

Those of us choosing this life understand the difference between Homeless and Houseless. We understand that it is a personal decision and not a last resort. We get the difference between "...a van down by the river.", and a van down by any damn river I choose (or mountain, beach, forest, etc.). We get it, but not everyone does.

Like it or not, you'll be fighting that stereotype and those prejudices for as long as you choose the Nomadic Lifestyle.

Why?

Because there is a fine line between a Professional Digital Nomad living on The Road and a homeless person living in a broken down RV beside The Road, especially to casual passers-by.

You've heard the phrase "Don't judge a book by the cover.", but make no mistake, your Rig is the cover by which you will be judged, every single day.

You have, no doubt, heard that some RV Parks and campgrounds don't allow Skoolies at all, and more and more cities are outlawing certain street parking and/or restricting the hours or types of vehicles allowed. - While this is more a result of the "scourge" mentioned above, we can't deny that we, as Nomads, bear a certain responsibility to the image we present.

If you're rolling around in a moldy RV, with a crappy paint job, belching smoke, you can't really blame the casual observer from making certain assumptions. - They don't understand the lifestyle. They have no point of reference. Why would they think any differently?

"But who cares what they think?" Right?

Well, you do. Specifically, you care what the cop or park ranger thinks when he or she is called out to have your Rig ticketed, towed, or just to threaten you until you leave (and this will happen). Even if you're not doing anything illegal, not making any noise, and just minding your own business, the reality of the situation is that your outward appearance will have a dramatic effect on how that conversation ends.

More importantly, how your vehicle looks and sounds is often the difference between whether or not that conversation ever takes place to begin with. If the neighbor or passer-by doesn't perceive you as a threat, if they don't think you're a freeloading drug dealer in a moldy RV who's going to leave behind a pile of stripped car parts and garbage, the phone call to the cop never even takes place, and he or she never shows up.

The best way to avoid, or at least minimize, all the prejudgments and bad vibes, be it from a small town cop or the old lady on the corner, is to keep your Bus or Van looking pretty, clean, and not sounding like it's about to explode.

Maintain Thy Rig.

While the aesthetics of your Bus, Van, or Shuttle are important, there are other types of maintenance that are an even more immediate concern to your daily life. Engine, drive-train, suspension, brakes, and staying on top of your fluid levels can mean the difference between being that Professional Nomad vs. Broken Down and Homeless as mentioned above. - It's one thing if other people don't understand. It's something else entirely if your are "dead in the water" because you forgot to check your coolant level.

It's another harsh reality to mention, but we are all just one breakdown and one bad day away from being stuck, and if you don't have an emergency fund or available parts for repairs, you can be stuck for a very long time.

You are going to learn to do a fair amount of your own maintenance, even if you have no prior mechanical experience. At the very least, you'll learn how to check your fluids and keep an eye on your tires. "YouTube University" can show you how to fix a surprising number of common issues (even on uncommon vehicles), and the library of Skoolie and Shuttle Bus forums can be a great resource as well.

Vanlifers have it slightly easier, due to familiarity and availability of parts, but at the end of the day, an engine is still an engine and brakes are still brakes. Sure, diesel engines and air-brakes add a few degrees of difficulty, but there is no reason not to at least be familiar with how they work and capable of doing some minor repairs.

You might consider yourself fairly handy with a wrench, or you may have no idea what words like "torque", "flywheel", or "differential" mean. Either way, keeping your Rig running is your responsibility. The days of waiting until the Engine Light comes on and pulling into the dealership are behind you, and you're Uncle Bob probably can't make the 500 mile journey to come save you from the side of the Road in rural Idaho.

In all seriousness, though, whether you're a novice, or even an accomplished shade-tree mechanic, here are few things that you want to do as soon as you possibly can, preferably before you even bring the Rig home:

1. Know your engine and transmission: - Find out the model numbers for both. Commit them to memory, write them down, save them to your phone, or etch them into the dash.

You should have this information before you even purchase the vehicle (it should be a condition of the sale), but if you don't, you can find out by looking for the number on a tag or stamped into the metal, do an online search using the VIN, or even take it down to the local mechanic (or diesel shop) and ask them to check. (Don't hesitate to pay them for their time. You're probably going to want to establish a good relationship with these folks.) You will need these numbers any time you order parts, look up how to do a repair, or chat with someone on the phone about whatever that "clunking sound" is.

2. Take yourself to Skool, er... School: - Once you have those numbers in hand, Google them. Do a search for the number followed by "issues", "repairs" or "forum" and start learning everything you can about them. Even if you plan on paying someone else to work on your Rig, this will at least give you an idea of what makes it go, what to expect, and some common terminology that will save you some time (and probably money) down the line.

If you haven't purchased a vehicle yet, or are uncertain what you're going to buy, try this experiment: Skoolie People, Google "DT466 issues". Read a few articles, watch a few videos and start getting to know how these engines work. Shuttle people should do the same, only substitute "6.8 LV10". Vanlifers have a lot more variables in engine sizes, so it might be better to try "Chevy Express Forum", "E 250 Forum", "Spinter/Transit/Promaster Forum" or something similar to get an idea. Even if you just spend an hour or two browsing the results, you will learn a great deal about maintenance and potential issues, perhaps even what to avoid when shopping.

3. Find the Vehicle Identification Number (VIN): - Yes, The VIN will be on the Title/Registration, and just like the Engine and Transmission, you will want to have this number readily available, as it is the key to several specifics that you'll need to know. In this case, however, we are talking about the physical location of the VIN on the vehicle. On a Van, or even a Shuttle, it will be on the dash, and likely on a sticker on the driver's side door, but in a Bus, there are many other possibilities, and you might need to find the hard-stamped VIN in the process of registering or transferring it to an RV Title (read the Tutorial at skooliesupply.com). Check for a stamped plate riveted into the metal above the driver's seat, look for it stamped into the firewall, or check the frame in the rear of the Bus, and/or under the side emergency exit.

4. Familiarize yourself with the specifics: - Once you have accumulated the above knowledge, the best thing you can do is find out everything there is to know about every part of your Bus, Van, or RV. There are too many individual items to list here, but to name a few: What size tires do you have? Does it have air brakes, disc, or drum? Is there a towing package? What's the capacity? What is the frame construction? Find out the gear ratio. Does it have a carrier bearing? Understand how the shocks, springs, and air bags work. Learn the preferred fluids (oil, coolant, transmission, power steering, etc.). What filters does it take? What belt or belts? Find out the alternator output.

Understanding these and so many more nuances about your Bus could be the make or break point in your Nomadic Journey. Keep digging through the forums and online groups. Talk to mechanics and other Nomads. Purchase the Chilton or Haynes Manual for it, if available. (You can find many Thomas and Bluebird Manuals online.) You should honestly become a downright nerd about it. - This is your home now, and the more you know about how it works, the better your life will be.

5. Establish a Maintenance Schedule: - This will look slightly different depending on the type of vehicle, but in short, you should research and establish a regular and set Preventative Maintenance Schedule. This means actually breaking out the tools, changing the oil, checking, topping, and/or flushing the other fluids, changing filters, changing belts, pulling the wheels and visually inspecting the brake pads/shoes/rotors, crawling under the vehicle to check the rear end, differential, axles, suspension, and frame, hooking up a meter and taking electrical readings, and much more.

Again, this is not an inclusive list. There are so many things will be specific to your type, make, and model, as well as issues that will be addressed at different times. You might change oil every 3000 miles, but only flush the radiator once a year. Brakes might last 20,000 or 50,000 depending on how (and what) you drive. Find out the specific recommendations for your Van, Bus, RV, or Shuttle, wright them down, commit them to memory, or have them tattooed, but stick to this schedule as if your life depends on it (because it kinda does).

Maintain Thy Rig.

Of course, maintenance doesn't stop in the engine compartment or lying on an oil soaked blanket staring up at the leaf springs. Remember that you have systems inside the Rig that you depend on to keep your life, and life on The Road going.

Electrical: - Once installed an functioning, you Solar Setup is relatively easy to maintain. Apart from regularly cleaning off your solar panels and making sure they are free of snow, mud, or any cracks and damage, the individual components don't require a lot of "hands on".

If you use lead-acid batteries you'll need to stay on top of the levels and make sure they are properly venting and functioning. It is also a good idea to visually inspect all of your wires and connections on a regular basis, keeping an eye out for evidence of shorting or excessive heat such as melted wires or burn marks. This can save you from costly repairs or even a fire later on.

The system itself will warn you of many potential issues. A breaker or GFI Outlet that keeps popping, fuses that regularly blow, or a repeated low voltage warning from your Inverter should be taken as signs of a serious issue and handled immediately.

The best thing you can do for maintaining your Electrical Setup is to invest in a decent monitoring system. They can be purchased for around $60, or as much as $400 on the top end, and they're relatively easy to install. You can see system performance in real time and note when things are less than optimal. If nothing else, use some of the 12v outlets that display voltage on them. A drop in voltage is also a sign of trouble and will let you know to take a closer look.

Water: - Maintaining your Water System will be a little more in depth, but it is absolutely vital that you do so. At the very least, you need to clean the holding tank once every six to eight months. Even if you go the bottled water route and only use this for cleaning, tanks can be a breeding ground for bacteria or mold. Let it go long enough and you can even smell it. The easiest way to do this is to run a little bleach through the system. About ¼ cup for every 50 gallons should do the trick. Use a funnel to pour it into your fresh water hose before hooking it up to the spigot (you're cleaning the hose as well). You then let the water push the bleach into the tank and fill it up all the way. We usually go for a quick drive to let it slosh around a little bit, run a few gallons through to tap to clean the pump and water heater, and then drain the tanks completely. Refill with clean water, run a couple more gallons through the tap to clear it out and you should be good.

Please note, if you're concerned about chlorine, go ahead and repeat the last step once or twice, but remember that there is very likely chlorine already in the water you're using to fill, especially if it came from a municipal source. Water test kits (or strips) are available on the Water Page at skooliesupply.com. They are inexpensive, easy to use, and great for piece of mind.

In addition to cleaning your Fresh Water Tank, you'll want to dump a little bleach in your Grey Tank as well. We usually just pour a quarter cup or so down the drain while we're doing the steps above. Since this is waste water there's no need to immediately drain the tank. Let it slosh around in there until the tank is full and drain as usual.

How often you change any inline filters will vary on the type, size, and how much water you use, but most manufacturers recommend changing every two to three months. Check the packaging for instructions. You should visually inspect the system for leaks or condensation on the lines and connections, and remember, if you're water pump runs when the tap isn't on, you have a leak somewhere. Track it down and repair it before it floods the Rig.

Propane: - The difficulty in maintaining your Propane setup is directly commensurate with how robust of a system you have in place. A simple set up like we described earlier in this book requires nothing more than checking levels of the bottles and occasionally replacing a connecting hose, or an o-ring or which is often as simple as swapping out the bottle itself.

Replacing the seal or O-ring on a propane tank is actually illegal in many areas, sometimes making the part difficult to find. Be sure you keep some on hand (they make a special tool for it as well). If you have a permanently mounted tank, you should be prepared to do this, or replace the tank entirely every few years.

For those of you with a full setup, there are a few more steps that you will need to take. Once a month or so you should spray down all the lines and connections with a soapy water mixture to check for leaks. If there are any bubbles, turn off the supply and repair immediately.

Maintenance certainly isn't the most fun part of this lifestyle. Keeping your Skoolie, Van, Shuttle, or RV looking good and running good can be time consuming and challenging, but it is worth every moment. It may not be as exciting as waking up to a beautiful vista or rolling into Quartzsite for the first time, but this is how you get there. This is your home now.

"Keep it runnin' like a top!"

Maintain Thy Rig.

Commandment # 6

Keep It Clean

"This is the worst pigsty I've ever seen!"

My mother, it would seem, was a renowned authority on pigstys, and my younger self and siblings were, apparently, over-achievers in the creation of such.

While maintaining the engine, drive-train, and systems in your Rig is an absolute necessity, maintaining the living space is equally as important, and some would argue even more vital to living a healthy life on The Road. Many new Nomads are downright shocked at the amount of time they spend cleaning, washing, wiping down, sweeping, dealing with trash, and even (you guessed it) human waste.

You might remember in the chapter about minding your resources where we mentioned a nice man in a noisy truck hauling away your trash, and we touched briefly on how everything in the Bus was your responsibility. Well, the biggest responsibility probably lies in keeping it clean. From dusting, sweeping, and getting rid of waste, to dishes, clothes, and every imaginable type of grime, your comfort, piece of mind, and even your health depend on a clean and well maintained environment.

You no doubt have a regular cleaning regimen in your sticks and bricks house, and most of what you face in your "rolling home" will be similar, however there are a few key differences and distinctions.

First and foremost, it is a much smaller space. Upon initial consideration, one might think that this will make it easier to clean, since there's less to do. Unfortunately, we humans, and our furry companions, create a certain amount of mess in a given day. It's like it's a set, static, or finite number. The amount of "dirtying things up" one person does is the same whether it's in a 2000 sq ft home, or a 60 sq ft Van. There's just less room to hold all that mess.

An old drinking buddy of mine used to ask for a shot of whiskey, then say something like, "Oh. I'm sorry. I'm feeling a little shaky today. Can you please pour that into a water glass?" The bartender would comply, and then, almost invariably, top off the shot with more whiskey. - It turns out that a legal shot of booze looks pathetic in a water glass, so my friend would slyly get extra hooch based on the appearance of his drink, even though the bartender had just poured it.

One "day's worth" of mess is sort of the same idea in reverse. When you have a whole house to hold the mess, it doesn't look like much. When you cram all that mess into a much smaller space, it's a lot more prevalent, and needs to be addressed more often. Add to that the fact that you (and your dog) are probably going to be spending more time outside, and tracking dirt, sand, and bugs into the Rig at a much greater rate than you likely do in your current 3 bedroom bungalow.

More mess + less space = more cleaning.

Additionally, there are a few cleaning chores that you will have to handle in a different manner. You'll no longer have the convenience of the guy in the garbage truck, sewer lines, or exterminators, and the small space creates a few unique issues of its own. Life on The Road requires a little extra effort in the area of hygiene, not just for yourself and your pets, but for the Rig as well.

Keep It Clean.

Since they tend to be the most unpleasant, we'll address the refuse we, ourselves, create first.

Trash: - The easiest thing to do with trash is to make a little deposit every time you stop at a gas station, rest area, grocery store, or state park (anywhere there's a trash can). You simply stuff a small bag or two into the refuse bin and casually walk away as if it was the most normal thing in the world.

This topic can present a bit of an ethical dilemma, especially if you're a large family, or generate quite a bit of rubbish. Garbage service isn't cheap, especially on the commercial level, and the cans are placed there as a courtesy to their customers. They are not really intended for household use.

One or two grocery sacs full of garbage (or better yet, bio-degradable bags), is okay, but please don't completely fill the dumpster at a rest area or try to cram a 33 gallon Hefty into the can in front of Walmart. When someone displays that kind of careless disregard it reflects negatively on all of us.

Be discreet, be reasonable, and be respectful.

The same goes for the garbage facilities at campgrounds. They are there for paying guests, but even if you did pay to spend the night, don't go tossing 3 full size lawn and garden bags in there. It's just bad form. We've even known people who've used dumpsters in an industrial park or behind a restaurant. - Same deal. These folks pay a lot of money for this service (and they probably have security cameras). - Spread your trash out into smaller bags and make more stops. Don't draw unnecessary attention to yourself, or to the rest of us. - It's a matter of class, common, sense and common courtesy. We'll touch on this again in an upcoming chapter about not being a dick.

It can also be surprisingly difficult to recycle on The Road. Unfortunately, while trash cans are relatively plentiful, recycle bins tend to be few and far between. They're not impossible to find, mind you, it just takes a little extra effort. If you're in a medium size city, a web search for "recycle bins near me" should bring up results. Certain grocery stores provide a bank of bins as a public service, and many recycling companies will allow drop off at their yard. Some even provide a means for depositing after hours.

There are a few apps that will help locate drop points ("iRecycle", for example), but if you're in the mountains, a small town, or boondocking, you should be prepared to pack it out. This presents a special difficulty for those with limited space, but bear in mind that it can be transported outside the vehicle. For example, a spare tire trash-caddy can be used. As the name suggests, it's a bag that straps to a spare tire cover, just watch out for bears and raccoons. Some Skoolie-Dwellers dedicate an under bus storage compartment, and we've known more than one who've used a roof rack Thule or Yakima bin for this purpose. Just be sure you thoroughly clean anything you plan to haul around for a while, seal it as best you can, and keep an eye out for those three arrows in a circle.

Toilet: - Welcome to the wonderful world of pooping in a bucket! - Whether you spend a thousand dollars on a compost toilet, a couple hundred on a cassette unit, or use a DIY Homer Bucket and a shovel to stir it, chances are you *will be* pooping in a bucket. - The good news, is that it's really not that bad, in fact, after a month or two it will seem perfectly normal, even relaxing. - Seriously! Grab a newspaper and enjoy the ride.

The three most common types of Nomad toilets are described below. The key difference is in cost, how you use them, and what you do with the waste. Remember, with no sewer lines, it's up to you.

Black Water: - Commonly found in factory RVs, black water simply means holding your waste in a tank, usually under the RV, and dumping it when the "shitter's full". For those of you with a black-water setup, you can ignore the rest of this section, download one of the many apps that show you black water dumping stations, invest in a good hose (and good gloves), do some research on what chemicals to use (it makes a big difference) and you're all set to go.

Cassette Toilet: - This is a portable unit that consists of typically 2 separate parts, the seat assembly and a 2 to 5 gallon holding tank. As with black-water, there is an air tight seal between the two, you do your business as usual (liquid and solid together) and then seal it back up. Some even come with a flushing mechanism. When it gets about half to ¾ full, you simply detach the bottom jug, walk it into a public restroom and dump it down the toilet. Rinse it out with a little dish soap, or vinegar and water solution, and you should have no issues with smells. A cassette toilet can last 2 people a couple of weeks before dumping, especially if you pee outside or in a jug, and try to use public restrooms whenever possible. If you spend most of your time in the city or at campgrounds or parks (regular options for dumping) a cassette toilet is a good choice.

Compost Toilet: - One quick misnomer to dispel up front: It is a "compost toilet" and not a "*composting* toilet". Composting takes months to accomplish, requires storage space, physically turning the pile, and often employs worms or bugs to aid in the process. - Not things you want in your Van. A compost toilet simply uses a "medium" such a coco coir, peat moss, or even sawdust to eliminate the smell until it can be disposed of.

It is absolutely vital to separate the solids and the liquids for the medium to work, which is why they come with a separate liquids container. If you pee in the solids side of a compost toilet it *will* stink.

As with the cassette toilet you can clean with vinegar and water, dish soap, or one of the many compost toilet cleaner products available. It is very important, however, that you don't use bleach to clean a compost toilet, as it can break down the enzymes in the medium.

But where (and when) do we um... dump it?

A family of two can typically go 3 weeks or more between emptying the solids tank of their compost toilet, and essentially there are 3 choices with what to do with the waste:

1. Bury it.: - Environmentally speaking, digging a hole is the best solution. Bear in mind that what you pull out of your toilet is not garden-ready compost. It still needs several months to break down. Short of a compost pit that is being actively worked, buried underground is the best place to do that. Sadly, while burying your waste quite doable when you're boondocking, it is rarely an option in town or in most public areas.

2. Drop it in a designated compost pile at a campground or park.: - The compost pile at a park is the best of both worlds. It is a pit that is being actively worked, and (presumably) the waste will be processed and used properly. Unfortunately, they're not always easy to find, so a lot of people don't consider this as a viable option. - Please note: Do not try dumping it with the black-water or directly into a stand-alone compost*ing* toilet, say at a trail-head or in a national park. In addition to the risk of contaminants (chemicals, baby wipes), there's the fact that you're adding to the workload of the, often underpaid, park personnel that have to service these things. Also, bear in mind that dumping compost directly on the ground (not in a designated compost pit) is illegal in almost every state. If there isn't a sign that specifically says "Compost Waste Here.", don't do it.

3. <u>Line the solids container with a biodegradable bag and drop it in a dumpster.</u>: - Many, if not most, people wind up going this route, just because it's more convenient than the first two options. While this may seem counter-intuitive, there are several websites that claim that it is environmentally sound. I'm not a huge fan of this method, personally, but I am often reminded, that people dump kitty litter and dog poop in the trash every day. If nothing else, as it breaks down, it can't be much worse than what's already in the landfill.

Pro Tip for the Guys: - The liquid container of a compost toilet can fill up really fast (especially after a few beers). I use an old laundry detergent container for my number 1. Not only will this add more time between dumps of compost, cassette, or even back-water tanks, but you can still stand when you pee. As with the jug on the cassette, you simply dump it when it's full, rinse out with vinegar or dish soap, and all is well.

This is another area in which you need to choose your battles. While compost toilets are a very visible part of the lifestyle, featured on several Skoolie and Van tour videos, they are expensive, require a fair amount of effort, and actually may not be the best choice from an environmental standpoint. Unless you can always dig a hole or find an active compost pile, a cassette toilet may actually be a better solution. Yes, you are still using water to flush, but we're talking about 1 gallon every 20 times you go, coupled with the knowledge that the waste will be leached into the ground or processed at the local sewage treatment plant.

They're a several videos and several opinions out there. Do some research and make your own choice, but from a standpoint of "keeping it simple" a cassette may be a good way to start. - Having done all three, a Cassette toilet is our current choice. We just try to No. 2 in public restrooms whenever possible. - You can always upgrade to (or DIY) a compost toilet later, once you have a better understanding of your needs and habits.

Keep It Clean.

Mold: - Moisture is a constant concern, and the accumulation of mold is one of the most common issues in Skoolies, Vans, and RVs.

When you introduce moisture into a small space, especially when there is a temperature difference between inside and outside, condensation will happen. We see it, mostly on the windows and perhaps counter tops, but it is, in fact, on every surface, and even penetrates wood grain.

What causes all the moisture? Well, we do. While propane appliances and heaters are a great source of condensation, the majority of the moisture in your Rig actually comes from the air that human beings exhale.

This condensation, if left unchecked, will create mold, which can not only a serious health risk, but can damage surfaces, and even the structure of your vehicle.

Since not breathing isn't a great option, here are a few ways to limit or reduce condensation, and thereby prevent, or at least minimize, the risk of mold:

1. Holes, slats, and mattresses: - If you've ever wondered why you see bed frames constructed with multiple holes or out of wooden slats in Skoolie videos, this is it. Simply giving the air somewhere to go prevents mold from forming. As does regularly airing out the surface by flipping the mattress and wiping down the frame.

2. Vents and fans.: - Keeping the air circulating in your Rig is the best way to let limit the risk. This is why reversible fans are a must have. We typically run one on intake and one on exhaust (even in winter) to keep air flowing.

3. Dehumidifiers.: What they do is self-explanatory, but a lot of folks don't realize that there are several low voltage, no voltage, and rechargeable dehumidifiers on the market. Many of them are inexpensive, and they make a huge difference.

4. Wipe down condensation.: - Whenever you see that moisture collecting on windows or counters, take a dry rag or towel, wipe it off and ring out the towel in the sink (or outside). The more you can send down the drain, or even what is being held in the rag, will make it less likely for mold to form.

Obviously, the best thing you do is to open all the doors and windows on a warm day, or run dry sources of heat (a diesel heater or even the onboard defroster, for example). Once mold is present a simple solution of 1 cup bleach to one gallon of water seems to be the most effective, but it is absolutely necessary to be ever-vigilant at keeping moisture and mold at bay.

Unwanted Visitors: - The presence of various vermin is often surprising to new Nomads. Ants, fruit flies, mice, and even rats are not strangers to Skoolie or Vanlife, and can get in to your Rig through the smallest of openings. Quite often the holes cut for plumbing, electrical or propane are the culprits. Seal these with spray foam or use steel wool and caulking to deter the little buggers.

Once present, another good tip is to clean their trail. Insects and mice usually leave a trail of pheromones to and from a food source. Spraying and wiping it down with a bleach solution, or even Windex (ammonia) can keep them from following their brothers and sisters to a free meal. Mint oil works well also.

You can also make efforts to prevent them from ever gaining access to your vehicle at all. A good move is to create a barrier with a product such as Home Defense or a similar insecticide. Whenever you're parked you can spray down the tires, levelers, or anything else touching the ground outside. We have found this to be very effective. - One note: If you travel with dogs or cats, go into the use of any chemical deterrents carefully. Start small and make sure your pets don't have a bad reaction.

The best solution, of course, is to not give the vermin a reason to stop by for a visit to begin with.

Every time you eat you do the dishes.... Right Then... Forget filling that bowl up with water and leaving it in the sink. Don't leave open cans of anything laying around, and you should even rinse any garbage free of foodstuffs before tossing it in the bag. Sweep your floor multiple times daily, and immediately clean up anything that spills. You'll also want to wipe down your counter tops and table, using your preferred disinfectant, after every meal and meal prep.

If you travel with children (even the adult kind) make sue that they don't leave candy or snacks anywhere, as the ants will be on their way before it even hits the floor.

Essentially, you're going to clean your Bus, Van, or RV to food service standards, be vigilant about how and where everything is stored, and regularly check for signs of unwanted visitors.

Keep It Clean.

Pets: - We all love our fur-babies, and most of them adapt very well to life on the Road. Dogs love it, and even cats can become quite comfortable. We've known Nomads with reptiles, birds, fish, rodents, a few spiders, and even a gal that travels with a dozen chickens and does great! The key is to take extra care with your pets, not only in cleaning up after them, but in managing their physical and emotional health.

All of the statements above regarding space, messes, and removing waste apply to your animals as well. They just don't quite understand the nuance of it all. Your cat is still going to track litter from the sandbox, and your dog is still going to dig in there looking for treasure. Animal hair will be everywhere. (Again, it's a smaller space, but the same size dog.) There will be several new rules and commands that you will develop in their new rolling home, everything from where they sleep to when they're allowed outside.

Be sure you plan for them to have a "place". They need their own safe haven to run to when they're nervous or sick, and you need to be prepared for that place to change. Take them on a few trial runs and make sure that they don't just want to lay on the couch or bed before you build that fancy cat enclosure. (Our cat typically sleeps on the dash.)

Like you, they will be spending a lot of time outdoors, and every time they come back in you need to be ready for the mess. It's a great idea to have a means of rinsing them off outside. Clean them up, wipe them down, brush them out, and check for fleas and tics. Disposable or washable floor mats will be your (second) best friend, and you'll both be thankful for a good outdoor rug for them to stretch out on.

Make sure you store their food and treats in an airtight sealable container (bugs and mice), and clean up any accidents the moment you find them. You'll want to regularly wash their blankets and beds as several parasites are invisible to the human eye. Watch to see if they scratch or seem uncomfortable, and plan on extra trips to the vet.

If you don't use to a national veterinary service, make finding a local pet hospital one of the first things you do when arriving at a new location.

Since all animals are slightly different, and come from different backgrounds, we strongly suggest that you have a consultation with your current vet, before you hit The Road, to discuss the specific needs of your pet. Remember, they're Nomads too.

Virus and Bacteria: - Also being more concentrated in a small space are the invisible monsters that make you and I sick. It's the reason cruise ships are quick to quarantine, and why people often fall ill after riding on airplanes. Keeping the air flowing and surfaces wiped down will aid in this realm as well as with mold, moisture, and pests, but you'll need to take special care in keeping yourself clean as well.

The cards are stacked against you here, because the very nature of living in a Skoolie, Van, or RV is that showers tend to be a little more sporadic. There is no reason, however, that you can't be clean and stay clean if you are willing to make a few adjustments and follow a few simple rules. We discussed "Navy Showers" in the chapter about minding resources, as well as joining a gym, but here some other means of managing your personal hygiene while using minimal or none of your own water:

1. Wash your hands: - We all had this drilled into our heads during Covid, and there's a reason. Your hands touch everything, your face, your food, and the door handle on the bathroom. It is the most efficient way for a virus to get into your system. You literally should take every chance you get to hit the hand soap or the sanitizer.

2. Change your clothes: - A lot of us wear the same shirt or pants 2 or 3 days in a row, but in sticks and bricks that's probably with a shower every day. If you don't shower *or* change clothes you're increasing the odds of infection by not giving the dirt anywhere to go. At the very least, change your socks and underwear (eww).

3. <u>Shower in the sink:</u> - Having spent some time as a truck driver, I have bathed in a rest area or truck stop bathroom more times than I care to remember, but the process is simple and effective. Get a small bucket or bowl full of warm water and two washcloths, use one to put soap on and another to wipe it off. Rinse and wring it out under the tap. Sometimes the floors can be a little nasty, so I typically bring in a towel to stand on, and a second one to dry myself. You can, of course, also do this at the sink of your bus and not have to worry about the nasty floors (or hiding in the stall if you're shy).

4. <u>Baby wipe bath:</u> - These are a decent way to clean up on those in-between-shower days. Washing yourself with baby wipes is like using hand sanitizer all over your body. You're not doing much for the dirt, but you are killing bacteria and viruses. - They do make a slightly larger "disposable washcloth" version, designed for this purpose. You can typically find them in the feminine hygiene section of most grocery stores.

There are several ways to get creative beyond these. If you're somewhere warm, you can just go swimming with a bar of soap. Use outdoor showers at the beach, pay admission to a public pool, or even visit the local YMCA.

These are just a few of the most common adaptations to life and cleanliness on The Road. You will discover many more, and while it may seem tiresome at first, you will adapt and develop a routine. Staying clean and staying healthy will make your travels so much more enjoyable.

"Pigsty? What pigsty?"

Keep It Clean.

Commandment # 7

Leave No Trace

"We were never here."

Life on The Road allows us to experience some amazing and beautiful things. There are so many sights, sounds, and vistas that not so long ago would have been nothing more than a dream, or a picture on a postcard, for many of us. The natural world, and how much time we get to enjoy and appreciate the wonder and beauty of nature is one of the main reasons we choose this life.

Be honest with yourself, when you daydream of being a Nomad, you envision watching the sun set over a lake or a a beach. You see vast mountains or endless desert. You imagine looking up at a clear night sky or into the deep blue sea. You dream of hiking in the woods, swimming in the stream, and long, winding roads that lead to beautiful places and picturesque panoramas. Mother Nature calls to you, and you immerse yourself in all the wonder she has to offer.

Even if it isn't your primary passion, the great outdoors is a huge part of this lifestyle. If you think of yourself as a homebody or a shut-in you might want to consider another plan. - If you don't already love being outside, you're going to figure out how to enjoy it (or you'll park the Van and go home).

We are so very fortunate to be alive at a time when what we are doing is actually possible. All of those rivers, mountains, forests, and beaches are truly a gift, but they are a gift that so many take for granted. So much of the natural world is already lost. It's been trampled under by over-use or surrendered to development. Reservoirs are drying up and the wilderness is being paved over. Despite an ever-growing interest in protecting the environment, there are still many areas that are threatened. We can't be certain if those classic landmarks, tourist destinations, and pristine places will survive our own lifetimes, let alone that of our children and grand children.

The natural beauty we get to experience is not a given. It isn't a right, and it's not a guarantee. If we want to continue to enjoy this amazing globe that we call home, it is our solemn duty to love, respect, and preserve that beauty. Those of us that are here now to appreciate it owe it ourselves, and to all of those who will follow in the future, to minimize our impact and protect it in any way we can.

Leave No Trace.

Up to this point in the book, we've been discussing ways to adjust and adapt to life on The Road, as well as practical advice for surviving, and even enhancing the journey. Hopefully we've given you some food for thought on effective ways to manage your Build, your resources, your living space, and perhaps even put your mind at ease at how attainable this lifestyle can actually be.

Ultimately, if you can find a balance between "what you want, what you have, and what you need", you'll do just fine. The key to success is to make simplicity and common sense your ever-present companions, and don't think everything has to be "perfect".

From this point forward, we're going to look at how we conduct ourselves on The Road, and what responsibility we bear to our fellow travelers, as well as the world around us. The last four "commandments" are going to address ethical considerations regarding how we interact with nature, animals, and other humans.

Please understand that we are not trying to promote any political or religious agenda, nor do we intend to scold anyone for their current or past behaviors. Think of these as "The Unwritten Code Of The Road". There are certain ways in which we need to interact, and certain rules that we need to follow. To fail to do so can have some fairly unpleasant results, but even as such, you're probably not going to "Nomad Hell", and the only penance you're likely to face will be the direct consequences of your actions, and very likely some unpleasant words from those around you.

Make no mistake, those consequences and unpleasant words can be extreme, but the bigger picture is about your own peace of mind. These ideas are meant to heighten your understanding and deepen your appreciation of the <u>ideals</u> of living as a full time Nomad. The closer you follow The Unwritten Code Of The Road, the better your life will be. The more thoughtful and considerate you are, the easier your path. - This applies equally to your mindset as it does your actions.

Several years ago I was at a campground on the Oregon Coast with a friend. We were relaxing after breakfast while the family in the next site was packing up their car to leave. We had our backs to them, but could hear the rustling of tents being taken down, the clanking of cooking gear, and the sound of water being dumped from the cooler.

The parents were shouting at the kids, "Roll up your sleeping bag!" "Where's the dog?!" No. You're not taking that stick home!" Their voices were loud enough that my friend and I rolled our eyes and shook our heads in annoyance more than once, increasingly anticipating their eventual departure.

We were both relieved to finally hear the sound of their engine firing up, and then subsequently fading off into the distance.

A few minutes later we noticed the unmistakable scent of plastic burning. I assumed someone nearby had just tossed a candy wrapper or something into their fire pit, but as the smell persisted and seemed to get stronger, my friend eventually got up to investigate.

"Jesus Christ." she muttered, with a tone of disbelief.

I rose to see that our family of loud-mouths had deposited a large bag full of garbage on their still burning fire, right before they drove away. As it sat there smoldering and stinking, my first thought was, *"How the hell do you even arrive at the conclusion that something like this is okay?"*

It was an unpleasant task, but with the help of the neighbors to the other side, we were able to extinguish the flames and pack the trash over to the dumpsters, which were maybe 100 feet down the road.

"You couldn't just shout at the kids one more time to take out the damn trash?"

Nope. They didn't bother.

We've all seen it before. In fact, it was commonplace for many years (at least in the US) for people to show a general disregard for the environment.

Many of us are old enough to remember a time before recycling, when there was no Earth Day, no Adopt A Highway Program, and almost everyone you knew would casually toss fast food bags and Styrofoam cups out the window of a moving car, without giving it a second thought. - You might even remember the early anti-littering campaigns featuring a Native American shedding a tear as a passing family through garbage at his feet. - Clearly our campground trash-burners missed that one.

As a kid, I once got paid to help a neighbor haul off a bunch of broken down washers and dryers that had been gathering rust on his porch. Did we take them to the dump or a metal recycler? No... We simply drove out a country road and tossed them down the riverbank. I don't believe that the option of taking them to the local landfill was even discussed.

River Road, near my hometown (the actual name of a country road that, not surprisingly, runs next to to the river), must have had a dozen such random dumping spots. It was common to see huge piles of rubbish that had accumulated over the years. Appliances, household garbage, even body-parts of old cars were simply pitched down the bank, as if it were nothing at all.

My mother used to take us to pick through them for random things of value like old signs, toys, and antique bottles. We'd walk along the river and dig through the junk as a pastime. It was an odd mix of appreciating nature, treasure hunting, and getting a taste of Americana (a "bad" taste, though it might have been).

In 1967, Arlo Guthrie released the anti-war anthem "Alice's Restaurant". - In the song, Arlo gets busted for dumping garbage next to the road, and that, ironically, helps him avoid the Vietnam Draft. Even as such, he makes light of it, as if it were no big deal. *"We figured that one big pile was better than two little piles, and rather than bring that one up we decided to throw ours down..."* - It just goes to prove that, not that long ago, even hippies were litterbugs.

It wasn't always like that. From the dawn of human history up until roughly the mid 1800s we reused re-purposed and recycled almost everything, out of necessity. The word "disposable" had very little meaning. The tools you used and the clothes you wore were made by hand and built to last. When something broke, you fixed it. When it could no longer be fixed you turned it into something else. Almost everyone fed and clothed their family by means of the local environment or working the land. The local ecology was respected and revered.

In a sense, we didn't need an Earth Day, because everyone was already doing it. Mother Earth took care of you, so you took care of her.

Over the last 200 years we've slowly devolved into a disposable society. Rubber tires replaced wooden wheels. Metal cans replaced clay pots and wooden barrels. Synthetic fibers made clothing cheaper to throw away than to repair. Plastic bottles replaced glass, and filtered cigarettes replaced cigars. By the mid 1900s there was trash everywhere, city and country. It was on every street and along side every road. We really didn't start seriously doing anything about it until the late 1970s.

The concept, "Leave no trace" isn't new, it just needed to be reborn.

It was also around mid century that outdoor recreation began to see an explosion in popularity. Inexpensive lightweight (synthetic) tents and sleeping bags, as well as the post war boom and the opening of interstate highways, made camping in previously remote areas more easily accessible. The National Park Service reported attendance surging from 33 million in 1950 to 172 million by 1970. People were coming in droves, and they were leaving their garbage behind.

You've certainly heard the expression "Leave no trace." before, but few realize that it originated as a training program. It was developed for employees of the National Park Service, US Forest Service, and The BLM, designed to help educate tourists on minimal impact camping, and general conservation. It was entitled "Leave No Trace Land Ethics", and it was so vital to wilderness preservation that it eventually developed into a full national education program with the name shortened to "Leave No Trace", alongside such memorable characters as Smokey The Bear and Woodsy Owl.

"The Leave No Trace Center for Outdoor Ethics" is a non-profit, formed in 1994, to create educational resources for ecological responsibility in outdoor recreation. They have organized the ideals of "Leave No Trace" into seven principles. They were originally directed more at those who hike in and camp in remote areas, but the overall theme can be applied to those of us who drive in as well.

    1. Plan ahead and prepare: - Basically, know the regulations of the area you're going to visit (pets, parking, etc.), as well as any special concerns or issues (endangered species, for example). Be prepared for emergencies and know the local weather. Avoid peak usage times and travel in smaller groups. Repackage food so that only take what you need. Use maps, GPS, and communication (cell phones) rather than marking a trail or hanging signs. (The iconic paper plate stapled to a tree comes to mind.)

    2. Travel and camp on durable surfaces: - Stay on the trail. Use designated campsites and park on rock, gravel, or sand. Don't take shortcuts through the trees or trample vegetation. - Unimproved campsites are *found* and not made. - Don't cut down trees, hack away bushes, or clear paths. Essentially, don't alter what is already there. It should look exactly the same when you leave as it did when you arrived.

    3. Dispose of waste properly: - Pack it in. Pack it out! Even if it's trash someone else left (or dumped into their fire pit), pick it up and take it with you or dispose of properly. This includes food as well. Yes, that apple core is biodegradable and the raccoons will probably enjoy it, but if there isn't an apple tree already there, it doesn't belong. - We discussed disposing of human waste earlier, and digging a hole is fine, just make sure it is 200 feet from any water source (rivers, lakes, etc.). This also applies to any dish water, bathing water, or the contents of your Grey Tank.

    4. Leave what you find: - Don't take away or add anything. Preserve the past. Don't touch cultural or historical artifacts. Leave rocks, plants, and other natural objects as you found them. Let's face it, we all love rock-hounding, but at the end of the day, the right thing to do is leave that chunk of rose quartz behind. (You don't need the extra weight in your Bus anyway.) Don't build structures or dig trenches, and don't introduce non-native plants.

5. Minimize campfire impact: - Where fires are permitted, use established fire rings, fire pans, or mound fires. Keep it small. Only use down and dead wood from the ground that can be broken by hand (or pack in/pack out your own firewood). Burn all wood and coals to ash, put out campfires completely, then scatter or bury cool ashes. Avoid having a fire at all if you don't need one. Use a camp-stove for cooking and a lantern for light.

6. Respect wildlife: - Do not follow, approach, and never feed animals. Giving them human food is bad for their health, alters natural behaviors, and can make them aggressive toward, or dependent upon, people. - Make sure your foodstuffs (and trash) are locked away or otherwise inaccessible. Avoid wildlife during sensitive times such as mating, nesting, or raising young. - Perhaps most importantly, control your pets at all times. I know that Rover wants to run free in wild, but remember that by allowing him to do so, not only are you risking the damage the your pup might unwittingly do, but you are voluntarily making him part of the food chain. Predator for some. Prey for others.

7. Be considerate of other visitors: - Be courteous! Everyone is here to enjoy nature, not to hear you shout at the kids, crank the music, run your generator all night, or blow your harmonica (We're serious about the harmonica thing... Really.). Keep your distance and camp away from others, unless invited. Respect their space, privacy, and the quality of their experience. In short, "Don't be a dick." - See upcoming chapter.

Leave No Trace.

The seven principles, laid out above, may seem like common sense to many, a little dated to some, and over-the-top to others. You might be a hard-core environmentalist or you might think global warming is a myth, but wherever you fall on that spectrum, living this lifestyle comes with certain obligations. The only reason you can enjoy that social-media-ready, picturesque sunrise parked next to a pristine lake, is because the people before you didn't mess it up. You are obligated to not mess it up for the group that comes along tomorrow.

Even if you don't follow each of these guidelines to the letter (though you should), you need to have a heightened awareness of the impact you have on the world around you, and you should try to minimize it wherever possible. If you don't do it for the environment, at least do it for your fellow Travelers.

These principles apply, not only to wilderness camping and boondocking, but also to stealth camping, rest areas, and even parking lots. While you may not see the urgency of protecting the blacktop down at Walmart, to fail to do so puts all of our ability to live this lifestyle at risk. No one will attest to the beauty of a Home Depot or Casino parking lot, but we will all miss the beauty of being able to park there overnight, once that privilege is taken away.

In addition to the above, those of us living in converted vehicles should take responsibility for our Buses, Vans, and Shuttles, as well. Think of it as the eighth principle, specific to modern Nomads.

We've covered a lot of points on this earlier, but considering it from a standpoint of leaving no trace: Make sure your Skoolie, Van, or RV is self-contained. When stealth camping, or parking lot surfing, you shouldn't have to exit the vehicle at all. Poop, pee, cook, eat, and sleep inside. You may find power hookups, water, or a porta-john at one of these places, but don't just assume they are yours for the taking. Ask permission before you plug into a random outdoor outlet like it's free Shore Power. Be mindful of any fluids (oil, antifreeze) that may be leaking from your bus. Even a slow drip will leave a stain on the pavement and possibly draw unnecessary attention (to say nothing of poisoning the local ground water). Find it and fix it ASAP, and don't leave spare or junk parts on the ground when you go.

By the way, if you can't fix that leaky oil pan just yet, you can use a cookie sheet to catch the fluid before it hits the ground. Additionally, as mentioned before, maintain your Rig, and make sure it isn't an eyesore. If you look bad, we all look bad.

There is one area in which you <u>should</u> leave a trace. If you're over-nighting in a commercial parking lot, be sure you go inside and make a purchase. The trace you are leaving is a few dollars for a bar of soap or can of WD40. Buy some groceries, have dinner, or drop 20 bucks in a slot machine. This only the reason they let us park there to begin with. Frequent the establishment and treat the grounds like it's a national park. Parking lot surfing may not be the most romantic or glamorous part of our lives, but it is often necessary, and it's a privilege, not a right.

No matter where you camp or what you're camping in, common courtesy and respect should be the order of the day, whether your host knows that you're there or not. Even when stealth camping, parallel parked on a residential street, you should treat the neighborhood as if it were a national monument, because it matters to someone. This is someone's home.

Boondocking in the desert, pulled off next to the highway, or up some remote logging road in the Pacific Northwest, it doesn't matter if you are there for the beauty or just the convenience, your very presence, and how you conduct yourself, makes a difference. It is up to each of us to minimize our impact, and do our absolute best to leave things as we found them.

The world belongs to no one, and everyone at the same time.
"We were never here."
Leave No Trace.

Commandment # 8

Pay It Forward

"I've always depended on the kindness of strangers."

From the film "A Streetcar Named Desire", the above quote is often taken out of context. While one may question the actual "kindness" that the character in the movie receives, the line is often adapted to promote the idea of "being nice to others" in one way or another. It has been referenced in books, TV shows, horror films, and even video games, typically in some way suggesting a general attitude of goodwill. "Be kind to strangers." and so on.

"Pay It Forward", of course, is also the title of a movie, released in 2000, that has turned the phrase into something of a cliche. The loose concept in this film is that, when someone does you a favor, rather than paying it back, you do three favors for three other random people. Whether you're a fan of the movie or not, it did bring this idea into the forefront. Since it's release, you've probably heard more frequent anecdotes of people paying the tab for the person behind them at the drive through or coffee shop (which seems to be the most common occurrence of the idea). You may have even noticed more friends donating to charity or engaging in other philanthropic activities such as volunteering or cleaning up garbage along the highway.

Beyond this, there is actually an entire foundation dedicated to random acts of kindness (randomactsofkindness.org) that suggests simple niceties that anyone can do, such as shopping for the elderly, donating clothing to a shelter, or leaving an unopened pack of wipes at a baby changing station. As with the movie, the thrust is to try to make the world a better place by through "good samaritan-ism", or simply doing nice things for others.

Paying it forward is a downright necessity when living on The Road. One of the scariest elements of this lifestyle is that you are out here alone, with no help, no backup, and few alternatives.

When you pull into that rest area in rural Alabama to sleep for the night, you are on your own. Help might be miles and miles away. If something goes wrong, you have yourself, your dog, and hopefully the good will of that guy over there in the Frieghtliner to help you survive.

Anyone can find themselves in a difficult situation, and make no mistake, it will happen to you. You will, almost certainly, at some point, give a fellow Nomad assistance when they need it, and you will, without doubt, accept assistance from a random stranger when you, yourself, are in need.

No matter how prepared you may be or how secure you think you are, you will, someday require the help of someone you've never met, simply to survive the day. Be it a ride to the gas station, or a tow to the next corner, we all depend, in some way, on the kindness of strangers, and Paying it forward is a way of life.

You can think of it as Karma, good will, or just a general belief in doing right thing, but true Nomads, <u>good</u> Nomads, will always lend a hand to those in need.

Last Thanksgiving, my wife and I parked on the beach and ate fresh crab at sunset. It was warm, serene, romantic, and perfect. We boiled our crab in fresh seawater. We dipped it in butter from a local dairy, and drank wine from a vineyard down the road. It was the best Thanksgiving ever. That is, until we went to leave and got stuck in the sand. To clarify, we didn't just get stuck, we buried the Rig to the axles. We dug and shoveled, pushed and pulled, and even offered to pay some rednecks to help get us out, but they drove right by us as if we weren't even there. To add to our pain and embarrassment, where we had gotten stuck was only about 20 feet from a paved surface that would have meant our freedom.

Now turning dark, and with the tide coming in, we had few options.

After several phone calls and many declines, we finally found a tow company that would send a driver out to "try" to save us from our sandy grave (at a holiday premium rate, of course). He estimated that it would be at least three hours before he could get to us. Perhaps he was the only tow-driver in the area that was working the holiday, or maybe he just needed one more round of stuffing and cranberry sauce before he fired up the ol' hook and headed South. (Either way, it was Turkey Day. - Can't really blame the guy.)

So with little else to do, we turned on the flashers and continued to shovel sand, in the hopes that, by some miracle, we'd manage to make our way out before the ocean dragged us in.

Our miracle came in the form of a couple of hippies that came strolling down the beach about 20 minutes later. Having noticed our four-ways, they had hiked nearly a mile down from a house on the cliff overlooking the ocean, for the sole purpose of rendering aid if needed. - Seriously, still full from Thanksgiving dinner, they came all the way down to where we sat, dead in the... sand... just to see if someone needed rescuing.

Let that sink in for a minute. That's what these folks did, on a holiday weekend... They were lying there feeling fat and stuffed, happy from turkey and mashed potatoes, but still decided to randomly make the hike, after seeing our flashers in the distance, just in case someone was in distress.

After the brief and obligatory, "Yeah, we're stuck. I can't believe this.... yadda yadda" discussion, my heart began to rise when one of them uttered the words "I've got a 40 foot school bus that can get to right here..." gesturing to the pavement that was so painfully close. - It was a fellow Nomad, there to help.

Forty-five minutes later (after they hiked a mile back up the hill to get the Bus), and with the assistance of 30 feet of chain (that they happened to carry with them on the Bus) we were on solid ground and back in business. They even helped us air up the tires that we had deflated in an effort to gain traction.

We had plenty to be thankful for. -It truly was the best Thanksgiving ever!

If you ever run into the Llama Bus or The LlamaBusFamily (follow them on Instagram), be sure to give them a big thank you from us, and purchase some of their really cool hand-made clothing.

Pay It Forward.

It goes without saying that anytime you see someone in distress, you should help them out. This could be as simple as a jump start or a bottle of water on a hot day, or it could mean physically helping repair a broken down vehicle, lending money, or even pulling someone out of the sand. ;)

Living this lifestyle is a dream come true, and we all can make the dream better, safer, and more rewarding for everyone.

"I've always depended on the kindness of strangers."

Pay It Forward.

Commandment # 9

Thou Shalt Not Be A Dick

"Everywhere I go there's an asshole."

When my wife and I were brainstorming ideas for this book, several potential Commandments were tossed around: "Honor The Bus Next Door", "Respect The Privacy Of Others", "Mindeth Thy Own Business", "Runneth Not Thy Generator Until The Wee Hours Of The Morning" and so on. Some were a little more vague; "Watch Your Height" (meaning "watch those low bridges" as well as "don't get to big for your britches"), "Mind Thy Fur Babies" (both for their protection and the protection of others), and a personal favorite, "Don't Shit Where You Eat" (literally and figuratively).

While the Commandments regarding how to set up your Build, and how to thrive and survive Nomadically were more or less obvious to us, those about how to conduct oneself on The Road were a little more difficult to phrase. It became clear, however, as we continued to write and develop ideas, that many of these potential Commandments, as well as several of the examples and anecdotes we've used in this book, could be wrapped up under a single heading:

Thou Shalt Not Be A Dick.

Not being a dick should be a remarkably simple thing to accomplish, yet it is, apparently, incredibly difficult for an unfortunate number of people in our society.

Early in life, we're introduced to The Golden Rule, "Do unto others as you would have them do unto you.", but we're shown, almost from birth, that success is often gained by doing the exact opposite.

A baby in its crib learns that it can manipulate its mother by crying. The louder you wail the more attention or food you will get.

By grade school we are actively teasing and picking on others to make ourselves look better. We laugh at the misfortune of our peers, and spread gossip and rumors meant to improve our status by undermining theirs. We steal each other's crayons and cut in line without giving it a second thought.

By the time we're teenagers, we're cutting people off in traffic, eating the last slice of pizza, and parking in the handicapped spot, completely indifferent, and righteously indignant as we're doing so.

As we reach adulthood "Dog Eat Dog" and "Cutthroat Society" become the order of the day. Success in business, or almost any other financial venture, seems inexorably tied to the exploitation of those of a "lesser" social status, or at very least, using, abusing, and confusing anyone we can, as long as it helps us achieve our own personal goals.

Buried deep within our reptilian brain is a reward center that makes it instinctive to do things that feel good, or get you what you want, without consideration of the needs of those around you. It's not a conscious thought. You just do it. The more you are rewarded for a certain behavior the more often you do it, and do it a greater extent.

We are rewarded for being a dick.

We're all familiar with The Golden Rule, and we've all heard those alternate versions of it, meant to be more accurate with real society. "Do unto others *before* they do unto you.", "Do unto others *or else* they'll do unto you.", or perhaps just, "Do unto others...". Look out for Number One and to hell with everybody else.

There are many of us who disagree with this behavior, often from a very young age. Long before we even understand what the word "empathy" means, we see people making fun of little Jimmy or little Susie and we feel sorry for them. We witness callous and inconsiderate behavior and can't understand why people do it (or think it's okay). Perhaps we've been the victim of such things before, and understand how hurtful it can be, or maybe we just don't understand how causing others pain benefits anyone else.

Doing the right thing, even for those of us that are empathetic (or just plain considerate), can be a struggle. We are often taunted or belittled when we try to be kind or stick up for someone else, and it's hard not to see how the bullies and bitches of the world tend to get what they want a lot more than the rest of us do.

Learning <u>not</u> to be a dick can be difficult, and not being a dick, in practice, can be laborious, if not outright painful. We are programmed to take advantage of the weak, and when we refuse to do so we, effectively, become "the weak", ourselves.

I grew up in an area where a lot of people were assholes, and fiercely proud of it. Every day I felt like a rubber ball, bouncing from one insult to another, one derogatory remark to the next, and downright abusive comments backed up by ignorant statements that everyone else seemed to accept without question.

My 7th grade Math Teacher told racist jokes in class, and no one batted an eye.

It was difficult for me to choose not to be a dick. I never wanted to be one, but it seemed almost a necessity to function in that society. It was even harder, in practice, to not be a dick. I was immersed in it. My interaction with almost everyone was abrasive by nature. A girlfriend once told me that I was a lamb in a sea of wolves, and while the comment was emasculating (and intended to be an insult), it wasn't entirely incorrect.

It wasn't until I moved to another part of the country, a place where being a jackass wasn't an expectation of "polite society", that I learned that this norm of dickery wasn't necessary. You weren't required to make fun of other people to advance your own personal agenda. You didn't have to take the credit when things were good, and pass the blame when they were bad. - Of course, you still had to exploit, use, and abuse people to get anywhere in the professional world. You just didn't have to be an asswipe after five o'clock.

It doesn't matter on which side of that coin you find yourself.

You might be fiercely proud to be an asshole, or a lamb in asshole's clothing, looking for a safe haven. If you are going to make it on The Road, you must choose to unlearn that dickish behavior which you've had pounded into your head your entire life.

Thou Shalt Not Be A Dick.

You might be thinking, *"I's not your damn place to tell me how to behave. I'll act however I want!"* (You just outed yourself.)

It's true, in the regular world, it isn't our place to be the ethics police, or to be all high and mighty about how people behave, but the reality of living Nomadically is this: The way you act and react with fellow travelers, and the way you interact with the environment, absolutely will affect your quality of life. It will come back to you...

Karma is real... That is to say, "Road Karma" is at least as real as the fact that your actions will have consequences. Even those things that don't seem to have an immediate effect on your day-to-day, can and will have long term effects on yourself, and on the rest of us.

Whether we're talking about a family that abandons their garbage in a still smoldering fire pit, someone who leaves derelict auto parts in the truck stop parking lot, or the guy two campsites down wailing on his harmonica until 3 o'clock in the morning, the optics of your actions affect how people think of Nomads in general.

People are much more likely to lump others into pigeonholes based on the bad behavior of a few than the good behavior of the majority.

We already struggle with how most folks don't understand the difference between "houseless" and "homeless". They think of us all as free-loading, drug-addicted, kleptomaniacs, leaving behind massive piles of garbage and debris all over the streets.

To the vast majority, we appear to be one in the same. More and more cities are passing overnight parking restrictions specifically geared toward preventing stealthcamping, ironically as a direct result of those who seemingly went out of their way to be as "unstealthy" as possible to begin with.

Many Walmarts and big box stores are restricting, if not completely prohibiting, sleeping in their parking lots. It doesn't matter of how nice your Rig is. They've dealt with enough indifferent and inconsiderate behavior, and now they're shutting us all down. It's not so much due to the homeless, but rather those Nomads who lacked common courtesy (as noted in the earlier chapter, "Leave No Trace").

Once, while we were sleeping in a restaurant parking lot, we were awakened by a group of partiers that had literally parked their vans and truck bed campers in a circle and built a bonfire in the middle. The music was blasting, dogs were barking, and raucous profanities filled the air. - Don't get me wrong. That stuff's all well and good, but do it on BLM land or in an empty field, not in the parking lot of a Cracker Barrel (especially the bonfire part).

When the cops showed up, they kicked us <u>all</u> out, including the Skoolie, Vans, and an RV that clearly had nothing to do with, nor had been a part of, the festivities. Even though we were on the other side of the lot, without so much as a cooler sitting outside of the Rig, we all had to go.

It's important to clarify that dickishness doesn't always have to be extreme examples, such as these. The simplest actions can kick your Road Karma into high gear. It can be as subtle as leaving your engine running all night, dumping your gray water on the ground, or peeing behind the bushes. Ignoring a no parking sign, blocking an entrance, or using all the paper towels in the restroom are all dick-moves. They're subtle, but still dickish. You may think it's perfectly acceptable to fire up the grill and have a cookout on the sidewalk, and while you may not hear anyone complaining, tail-gating is for football games, not out in front of the Home Depot in the middle of the day.

Then there are those people who just *look or act* overly "creepy". They park too close, they stare and ogle passers-by, or even panhandle customers as they come out of the store.

Remember "Mind Thy Fur Babies" from before? We shouldn't have to say this but, your pets don't understand that they're being dicks. - It's your responsibility to make sure they're not. - While I'm sure your dog is an absolute sweetheart, I still don't want him snooping around my campsite after dark or marking my tires when you're not watching.

While I, personally, love dogs and cats, many people don't. Some are actually terrified of dogs, many are allergic to cats, and pretty much no one wants a nose in their crotch or a hair in their food. Incessant barking at a rest area or scenic overlook? Nipping at children and stealing their ice cream? Attacking other dogs and cats, or starting fights? Dick. Dick. Dick. (See Dick be a dick.)

If you can't control your animals, then they need to stay in the Bus or on the leash, and if you don't *care* that they are annoying or frightening others, then you should stay on the leash as well. This is so unfortunately common that it almost warranted a chapter all its own, but as noted, it ultimately comes down to the same point.

Thou Shalt Not Be A Dick.

All of these things steadily reinforce the negative stereotype, and are usually downright abrasive to your fellow Nomads. Things that may seem trivial to you might be a big deal, or a major inconvenience, to someone else, perhaps even a big enough deal to lodge complaints or pass laws.

Ultimately, its your business what kind of a person you want to be, but how you conduct yourself in your daily affairs reflects on the entire Nomad Community, and we are already under a fair amount of scrutiny.

If you decide to be a jackass, I'm not going to stand in your way, but I'm also not going out of my way to do you any favors (like pull you out of the sand when you're stuck on the beach). If you're a big enough jackass, you might even risk retaliation, be it from a store manager, police officer, a fellow traveler, or even that guy over there in the Peterbuilt who's trying to get a little shut-eye before he has to log another 800 miles in the morning. It is a simple thing to be aware of people around you and exercise a little common courtesy. We're all in this together, and we're all out here on our own, at the same time.

"So..." You might be wondering, *"Am I a dick?"*

Well, if you've been rolling your eyes throughout this chapter, thinking things like, *"What? What's wrong with that?"* If you have no hesitation parking in the handicapped spot, blast your bass-thumping, window-rattling, stereo at the stop-light, and you still don't understand the importance of, at least, not acting like the many examples we've shared so far in this book, then there's a good that chance you are.

Our society has devolved, greatly, just over the last few years. It has become normalized for people to think only of what they, themselves, want, and have no regard for the needs of others. By definition, they're saying, "What's best for me is what's best, and screw the rest." Racism, misogyny, ignorance, and hatred have all become commonplace. Some folks are even vastly proud of the fact that they ignore the needs of other's. They use words like "snowflake" and "triggered", as if anyone who has an issue with their actions or words is somehow "less than" themselves. (Kinda reminds me of home.)

If that's you then, yes, you're a dick.

Whether you're a dick about respecting the places you park, respecting the people around you, it just doesn't occur to you that other folks have the same needs as yourself, or you feel like those needs are somehow less valid than your own, these are behaviors that you will have to change.

Put the toilet seat down and put the paper on the roll. Don't warm up your diesel engine for 30 minutes at five in the morning in a crowded campground. Don't drive like an ass, don't park like an ass, and don't show your ass to people who don't care to see it (literally or figuratively).

To be clear, you're welcome to think whatever you want, and you're welcome do whatever you want as long as it doesn't affect anyone else. You don't have to agree with someone else's feelings. You don't have to understand why someone might have different thoughts than yourself. You don't even have to actually give two shits about another person who might look, think, or pray differently than yourself. - You simply have to <u>not</u> be an asshole about it, and consider, for a moment, that you're not the only one that matters.

It's honestly true for life in general, let alone living on The Road. Your life will be better if you treat others better. It costs you nothing to refrain from pissing other people off.

Keep yourself in check, let others be, take responsibility for your Rig, your trash, and your animals. Be aware of and considerate about your surroundings. "Mind Your Height", and "Don't Shit Where You Eat".

All you really have to do is exercise a little common courtesy, and take a moment to consider how your actions help, hurt, or harm the rest of us. If you can't do that, at very least, try not to be a pain in everyone's ass. It's simple... Like, painfully simple... Yet, surprisingly difficult for a lot of folks to comprehend. - While you should "Pay It Forward", as we discussed in an earlier chapter, if you simply can't find it within yourself to do nice things, or make an effort to help people, you can, at least, *not go out of your way* to be a jackass.

"Everywhere I go there's an asshole."

Thou Shalt Not Be A Dick.

Commandment # 10

You Can Always Drive Away

"You don't have to attend every fight to which you are invited."

That sage advice was imparted to me by my grandfather when I was about 8 years old. Over the course of my childhood, he and I had participated in many profound conversations, but this particular one came on a Spring afternoon while I was working for him on the farm.

To be clear, I was working on the farm due to the fact that I'd been kicked out of school.

I'd gotten in a brawl with an older kid (a bully) who had been picking on myself and my friends, and while I, admittedly, had lost the fight, the action still warranted a three day "vacation" from classes, which, ironically, turned out be three of the most educational days of my life. Granddad was only too happy to have an extra set of hands with his daily chores, and he took the opportunity to pass on this and many more words of wisdom to his young grandson.

At first, of course, I thought he was referring only to physical fights, like the one that I was reminded of every time I tried to speak or eat, still sporting the fat lip that I had gained during recent entanglements. It wasn't until many years later that I came to realize that "not attending every fight" applied to verbal altercations as well. Arguments, shouting matches, or even an exchange of dirty looks with some idiot who doesn't know how to merge left into traffic. In fact, any disagreement that results in any kind of heated exchange, is something in which you are free not to participate.

You Can Always Drive Away.

I'm sure Grandpa actually said something like, "Walk away..." or perhaps, "Turn away..." or even more likely, "Ignore the bastard and don't let him waste your time..." but the spirit of the statement is the same. We, as Nomads, just have the extra advantage of being able to fire up the engine and hit The Road, rather than sticking around and dealing with somebody else's bullshit.

Growing up in a very small town (full of assholes), I didn't have that luxury. I literally knew everyone, and everyone knew all of everyone else's business. If you made a mistake, or said the wrong thing at a party, everyone knew about it before breakfast the next morning. My meaningless little schoolyard scuffle had even apparently been a topic of conversation at the local tavern.

"Heard you got into some action today, son." my father chided as he walked in the door that evening. I didn't even get a chance to tell him what had happened before he was on to showing me the finer points of self defense and how to effectively throw a left hook. - He not only knew that I had gotten in a fight, and that I'd lost the fight, but even how many punches the other kid had landed, compared to my two or three. Apparently the play-by-play had been repeated in every gathering place around town, including the bar where my dad typically stopped in for a couple of beers on his way home from work. - I was a small town. - I can't tell you how many times the fact that I'd lost a fight (when I was 8) came up in conversation over the next several years.

The point is that I couldn't just head for the horizon and pretend that getting my butt kicked on the playground had never happened. I had to stick around and take my verbal licks from friends and classmates, in addition to dealing with the embarrassment of a playground ass-whoopin' for years to come. I'm sure, to this day, that there are kids from the old hometown who still remember it.

Whether you are a full-time Van or Skoolie-Dweller, or just a weekend warrior who takes extended trips in the RV, you have the advantage being able to drive away. I'm not talking about tucking your tail and running, but simply not bothering with an abrasive situation, or abrasive individual. - It just isn't necessary. - Since you don't have to stick around and deal with the same idiots day in and day out for the rest of your life, there is very little value to actually bothering with the fight/argument/staring match to begin with. It is, almost always, of little or no consequence. All you'll gain for your time and effort is a bruised ego, elevated heart rate, or at best, the feeling that you are right and someone else is wrong (which is exactly what the other person will be thinking at the same time).

Make no mistake, you will run into assholes. People will try to start trouble and stir up drama. You'll be attacked for no reason or (most often) very aggressively be told that you can't park somewhere, when a simple "would you mind moving" would have sufficed. The beauty of Nomadic Living is that the folks yelling at you for no reason and the "dick" (from the last chapter) who is unreasonably "dickish" with seemingly no other cause than to advance the art "dick-dom" doesn't matter anymore. You literally *never have to see this person again,* so why waste your time and energy?

You Can Always Drive Away.

We are obsessed, in our society, with being "right" and winning the conversation. You'll see people debate utterly meaningless issues for hours on end, for apparently the sole purpose of claiming the moral or intellectual high-ground (or simply annoying everyone else at the party).

I once dated someone who would start arguments over trivialities for seemingly no other reason than to hear me utter the words, "You're right, honey, and I'm wrong." It was as if she couldn't get through the day without winning a fight, even if she had to go out of her way to start the very fight she needed to win...

Needless to say, the relationship didn't last long. - I drove away.

At the end of the day, if you seriously think about it, being "right" has no real value at all. You can argue with someone ad-nauseam, and when you're finished, you both, in reality, are just thinking that the other guy is an asshole (and that may be the only thing that you're both, actually, "right" about).

Sincerely, have you ever walked away from an argument thinking, "Gosh, they make a good point, I really should get my shit together and stop being such an idiot..."? - No... You haven't. It could about politics, religion, or even if it's legal to turn left on red from a two-way to a one-way street (it isn't, by the way). The debate, fight, or altercation almost never resolves the issue, and often just leaves you both worse off for the effort.

Forgive the misogyny, but something else my grandfather used to say is, "Arguing with a woman is like teaching a pig to dance. It's a waste of your time, and usually just pisses off the pig." - He was joking (sort of), but if you remove the gender-context the statement is surprisingly true. Arguing with <u>anyone</u> is like teaching a pig to dance...

When that random homeowner or park ranger comes at you, vigorously shaking a finger and shouting, "You can't be on this street!" or, "No overnight camping!" you might be tempted to stand up for yourself and reply in kind with, "I'm allowed to park here!" or, "I'm not breaking any laws!", and while that may be true, is it really worth the fight? You can argue your point until you're blue in the face, but even if you are, in fact, "right", I can assure you that they don't see it the same way (or they wouldn't have been wagging the finger from the start), and they will not rest until they have "won" this particular conversation. - This includes calling the local authorities, or by whatever other means, making your life a living hell until you comply with their demands, no matter how ridiculous or "wrong" they may be. - It's ultimately a whole lot easier to just go park somewhere else.

Think of it as a bad smell. If you were camping somewhere, and suddenly noticed a rank odor blowing in on the wind, the scent of a skunk, for example, would you argue with it? Would you try to "win", as if it were some olfactory competition? - No... You wouldn't. You would simply pull up stakes and go camp somewhere else. Spray as much perfume on it as you want, but there's no sense in having a fight with a smell, just like there's no sense in teaching a pig to dance.

The beauty of living Nomadically is that you can park/sleep/camp any-damn-where-you-please. You don't have to deal with the bad smell, the finger wagging homeowner, or the pig trying to learn the Tango. (You can always just do the Hokey Pokey and turn yourself about.) There are limitless other places you can be. Pick any one of those, go there instead, and focus your energy on better things.

You Can Always Drive Away.

I can't tell you how many times we have left (driven away from, or avoided) an unpleasant situation, only to find a much better alternative.

My wife was once driving our 40' Skoolie up I-35 when we got pulled over by a state trooper. She didn't have a Class B CDL, and the cop was convinced that she needed one. (See our tutorial on CDLs at skooliesupply.com). For the record, she <u>didn't</u> actually need a CDL to drive our bus with a Vermont RV title and seating for 6 (we were "right"), but the trooper wouldn't let it rest, even after we showed him that very tutorial, and clicked on the link to a website referencing the Texas State Laws on the matter.

It didn't take long to realize that there was no way out of the situation in which he didn't "win" the conversation. - Fortunately, I did (in his opinion) have the proper license, and because we hadn't been "difficult" about it, he agreed to let us continue to the state line without writing us up, as long as <u>I drove</u>, and my wife didn't.

Ultimately, for us, it was less important who was "right" than it was that we were allowed to continue. Being polite and avoiding the confrontation was the best/easiest resolution to the situation.

Props to my wife for being secure enough in her womanhood and letting that particular example of patriarchal BS slide, and also, props to her for the jokes later that evening about the cop's … endowment... (He clearly drove a very large and very lifted 4x4 in his personal life).

Being "right", or winning the argument, is just one of the many things that we have been conditioned to value while living in Sticks-And-Bricks society, and yes, in that realm it may have a certain worth. Just like being in a hurry, having the most toys, or the biggest, best, and latest gadgets, these are just a few of the ideas we can blissfully leave behind when living on The Road.

"You don't have to attend every fight to which you are invited."

You Can Always Drive Away.

The Journey Is The Destination

"Are we there yet?"

For the true Nomad, the answer to every parent's favorite question (above) is, "We were always there, but we will never be finished getting there..."

Some of the best advice that I ever received came from a crusty old Van-Dweller who was parked next to a gravel road in Rural Indiana. It was literally in the middle of nowhere. The closest house was about a mile away.

It was just after the corn had been harvested, so you could literally see flat and "nothing" for miles and miles in each direction. I was visiting a friend who lived nearby, and after discussing why someone might be randomly parked at such an odd place, we drove down the road to see if, perhaps, they were broken down, or in need of assistance.

When we arrived, he was sitting there with the sliding door open, having a beer, eating some Kentucky Fried Chicken, and staring out over a vast expanse of, now empty, fields as far as the eye could see. After a brief chat and confirming that he wasn't in any peril, my friend asked the obvious question, "Why did you choose this particular spot for a picnic?"

"The secret to living a good life..." the old man replied, "is learnin' to appreciate the places yer passing through while yer goin' where yer goin'."

The Journey Is The Destination.

The fact that you are choosing this lifestyle, means that you're, most likely, already of this mindset. You're done with the rat race and keeping up with the Joneses. The idea of appreciating the "space between the places" has probably been on your mind for quite some time. You decided, long before you ever started reading this book, that there was another way.

Make no mistake, breaking that cycle isn't easy. We've spent a lifetime "being somewhere", or needing to be somewhere at a certain time, or before a certain thing happened. "Billy, be home before sunset..." or "Your curfew is 10pm...", "Your shift starts at 5...". Our lives have been ruled by "where" and "when". School, work, dates, even "get me to the church on time..." have caused us anxiety, and cost us sleep since we first learned the difference between the big hand and the little hand back in kindergarten.

It is the reason so many retired people go back to work after a year or two. They feel like they have no purpose if they aren't "on the clock". Doing nothing is a lot harder than it sounds, and lots of folks find that *not* having to be anywhere from 9-5 is uncomfortable, if not outright awkward.

There will, of course, be situations in which you'll need or want to arrive on a certain schedule. "The park opens at 8..." The show starts at 9..", etc., but for most of the rest of your life, that estimated arrival time in the lower portion of your phone screen will be unnecessary, unimportant, and unworthy of your concern.

You're going to go to some amazing places and take in some breathtaking sites. You'll be able to visit destinations that you never thought you would, didn't think you could afford, or never even knew that you wanted to see in the first place. You'll sleep in places that should be postcards, attend events and festivals that you've dreamed of for years, and eat in restaurants that you've only seen on TV. The destinations are, for sure, destinations, but your new lifestyle will allow you do so much more than that.

Every side road, every detour, and every little town along the way is an opportunity for a new adventure. Take the exit. Go to the farmer's market. Explore the beach. Hike the trail. Go the long way. Open your Van door and have chicken and beer in a cornfield.

The Journey Is The Destination.

That's why You're here. That's why you started dreaming about this in the beginning.

Hopefully, in this book, we have given you some food for thought, and maybe even helped you avoid some expensive mistakes.

Perhaps you've come up with "Commandments" of your own, or at least have a better picture of what it takes to make it on The Road.

By no means should the tenets we've discussed here be considered the "only righteous path", but rather something you can adapt and adjust to your own lifestyle and situation.

For a quick recap, let's scroll through them once more. Consider how each one may affect you, and view it through your own lens.

You Are Never In A Hurry. It's time to smell the roses and count the ants. You are choosing time and serenity over drama and complications. Take the time to do what needs to be done, and give yourself the time to enjoy The Road, and all it has to offer. Learn to appreciate the slow lane. There is no "sense of urgency". You have all the time in the world. To have the time and not take the time would be the biggest sin.

Less Stuff = Less Stress. After a lifetime of "clutter", both literal and figurative, you've come to a space in which you long for more than just the latest toy's, gadgets, and goodies. You feel the weight of a lifetime's accumulation. Every day, from here on out, you will realize a little more, and a little more, how much it all holds you down. Consider the true cost and remember the Parable of Prepositions. How many times are you willing to move something out of the way. Free boats aren't free.

Mind Your Resources. Consumption is no longer your god. It's time to turn off the switch, shut off the tap, and consider what comes in, as well as what goes out. Taking control of your life, means taking responsibility for what you use, and understanding what you don't need. Rethink how you cook, eat, drink, and bathe. Unlearn the wasteful habits of the past and your Nomadic Voyage will be easier and better in so many ways.

Keep It Simple. You don't need a Ferrari when a Chevy will get you there. Your Build will never be perfect. Your relationship with your home on wheels will constantly change. What you think you need today may well become cumbersome further down The Road. Despite what you see on social media, you're not building a five-star hotel. If that's what you want, pony up the cash and stay at one. Day to day, on The Road, is about function more than form.

<u>Maintain Thy Rig.</u> Your Bus is your home. You might be one breakdown away from devastation. The days of just turning the key and dropping it in gear are gone. Everything from the oil to the oven, and the tires to the solar panels is your responsibility. It is a labor of love.

<u>Keep It Clean.</u> Mold is your new nemesis. Flip the mattress and air it out. You need to be up close and personal with every surface and every appliance. Look out for bugs, dust, dirt, grime, and pet hair. When the space is less, the greater the mess!

<u>Leave No Trace.</u> "We do not inherit The Earth from our ancestors. We borrow it from our children." This Native American proverb is your new mantra. You want the beauty to be there long after you're gone. You respect what you see and appreciate how fragile it all is. Stay on the path. Pack it in. Pack it out.

<u>Pay It Forward.</u> You never know when you're going to need a lifeline, but you can be certain that someday you will. Someone might save you from a would-be sandy grave, and you will someday lend a hand to someone in distress. We're all out here together, and we're all on our own at the same time. Give what you can. Be what you need.

<u>Thou Shalt Not Be A Dick.</u> It should be the first commandment for life in general. There's no need to be abrasive, indifferent, thoughtless, or just plain mean, and on The Road it's even more important. One careless overstep can have a huge impact on all of us. Consider the needs of others, and "Honor The Bus Next Door".

<u>You Can Always Drive Away.</u> When someone ignores the previous commandment, it's okay to ignore <u>them</u>. Having the fight, argument, or even just the stink-eye is rarely worth your time, and almost never ends well. You now have the luxury of putting them in your rear-view mirror and keeping it that way.

So many aspects of your life are about to change, from taking your time, to letting go and keeping it simple. You no longer have to "play the game" the way we've always done in the past. The old rules don't apply, and every day is an adventure.

Many people start the dream of living in a Skoolie, Van, Shuttle, Box Truck, RV, or even a Subaru, thinking that they have to find a way to re-create the life they've lived so far, when, in fact, the very idea, the very purpose of this lifestyle is to <u>not</u> do that. It is to let go of the demons of the past. It is to leave behind the expectations and obligations of corporate society, and change your outlook forever. It is to live free, and live on your own terms.

Living on The Road will be the best time of your life. It will be magical and awe-inspiring. Whether you spend the rest of your days behind the wheel or look back at it years from now as your greatest adventure, you, and your world, will never be the same.

"Are we there yet?"

The Journey Is The Destination.